# A Long Ride in Texas

NUMBER FIFTY-ONE
*The Centennial Series*
of the Association of Former Students,
Texas A&M University

# A Long Ride in Texas

.

THE EXPLORATIONS OF

JOHN LEONARD RIDDELL

EDITED, WITH AN INTRODUCTION, BY

## James O. Breeden

Texas A&M University Press
College Station

*Frontispiece*
John Leonard Riddell, circa 1855. Courtesy Rudolf Matas
History of Medicine Collection,
Health Sciences Center Library
Tulane University

The paper used in this book meets the minimum requirements
of the American National Standard for Permanence
of Paper for Printed Library Materials, Z39.48-1984.
Binding materials have been chosen for durability.

## Library of Congress Cataloging-in-Publication Data

A Long ride in Texas : the explorations of John Leonard Riddell /
edited, with an introduction by James O. Breeden — 1st ed.
    p.   cm. — (The Centennial series of the Association of Former
Students, Texas A&M University ; no. 51)
    Includes index.
    ISBN 0-89096-582-X
    1. Texas Hill Country (Tex.)—Discovery and exploration.
2. Botany—Texas—Texas Hill Country  3. Geology—Texas—Texas Hill
Country.  4. Riddell, John Leonard, 1807–1867—Diaries.
I. Breeden, James O.  II. Series.
F392.T47L66  1994
976.4—dc20                            93-27569
                                         CIP

For Chris and Pappy
TRAVELERS IN TEXAS

# Contents

vii

# Illustrations

# Preface

Early Texas witnessed a succession of travelers. Among this steady stream of visitors was a handful of scientists.[1] Almost exclusively students of the natural world, these natural historians, as early naturalists were called, penetrated every part of Texas. Dr. Ferdinand Roemer, the prominent German natural scientist and so-called "father of Texas geology," laid claim to being the first naturalist to visit the Hill Country of Central Texas in 1847. In reality, John Leonard Riddell of New Orleans preceded him by almost a decade.[2] A native of New York and educated at Rensselaer Polytechnic Institute, Riddell made his mark in science in New Orleans, where between 1836 and 1865 he was professor of chemistry in the Medical Department of the University of Louisiana (now Tulane University). Riddell's life in science was characterized by versatility. He contributed to botany, chemistry, geology, physics, and medicine. Thomas Cary Johnson, Jr., the first systematic student of science in the South, characterized Riddell as "an indefatigable observer and a bold and original thinker."[3] Riddell's international reputation and extensive bibliography—nearly eighty publications on a wide variety of topics—underscores this assessment.

Riddell made two trips to Texas, both in 1839. He spent some five weeks in the Trinity country of the southeastern portion of the state during April and May conducting scientific investigations. The results, a survey of the topography, geology, and economic resources of the region (reprinted in this volume), appeared in the prestigious *American Journal of Science and Arts*.[4] In September, he returned and stayed nearly three months as a scientific consultant for a company attempting to locate the fabled lost San Saba silver mine in west-central Texas. The subject of a popular Texas legend, this mine was reportedly discovered in the 1750s by Don Bernardo Miranda, the military commander of Texas, near the confluence of the Llano and Colorado rivers. A mission, San Sabá de la Santa Cruz, and fort, San Sabá Presidio, were established to enslave Indians to work the mine and to safeguard the operation. Years later, after a fortune in silver had been stockpiled in a treasure room deep within the mine, the Indians revolted against the severity of the Spaniards and killed everyone. The mine was sealed, except for one secret entrance. In the 1820s, Indians, whose friendship he had courted, showed Jim Bowie the secret entrance. He recruited a band of followers in San Antonio, made several trips to the mine, and removed a large amount of silver before the Indians turned on him and forced the group to flee. Since only Bowie knew the full secret about the mine's hidden entrance, its location died with him at the Alamo. While numerous people were subsequently said to have stumbled onto the secret entrance, no one was ever able to retrace his steps to the lost San Saba mine.[5]

Little is known about the origin and details of the 1839 venture, other than that Matthew Hopkins was its president and Riddell was an equal partner in it with nine or ten investors.[6] But he had put up no money, as the others had, to purchase land around the old Spanish fort. Rather, his contribution was to be his knowledge of geology and mineralogy. Ambitious and opportunistic, Riddell looked to the undertaking as a possible means to enrich himself. The exploring party reached the general locality of the mine but did little to locate it, seemingly from poor planning and fear of the Comanches. But all was not lost, as Riddell received for his service the equivalent to ten thousand acres of Texas lands. Moreover, the trip was successful from a scientific point of view, as Riddell made pioneering botanical and geological observations and brought back numerous specimens of Texas plants.[7] He was elated in the discovery of numerous previously unknown species. "Of a great number of new plants," he wrote in his personal journal, "I have the descriptions and fine specimens taken on the spot where they grow."[8]

Riddell kept a detailed diary—numbering ninety-four pages—of his travels in Texas. No sooner had he returned to New Orleans than he began to make plans to publish it. Even an appointment as melter and refiner for the United States Mint at New Orleans, which awaited him on his return, and the death of his first wife, a week later, did not deter him. One of Riddell's botanical correspondents, for whom he had previously collected plant specimens, was John Torrey of New York. A leading student of nature, Torrey was, in conjunction with Asa Gray, his botanical partner and the nation's premier botanist, preparing a comprehensive *Flora of North America* (1839–43). In February, 1840, Riddell wrote Torrey to inform him of his travels in Texas and to promise him any duplicate specimens he might have. In return, he asked: "Where I furnish you a satisfactory description of a new plant, in accordance with the plan of your flora, will you let it appear in your flora, in my name?"[9] At the same time, he wrote to Dr. R. M. Patterson, director of the United States Mint at Philadelphia and a trained scientist, to intercede on his behalf with the American Philosophical Society, of which Patterson was a member and later president, to publish his Texas plants with colored copperplate engravings.[10] Nothing seems to have come of the overture to Patterson, but

Torrey included at least ten of Riddell's specimens in his *Flora* and named a previously unknown species of *Senecio* in his honor.[11]

In July, Riddell wrote several New Orleans newspapers, proposing a biweekly series on his travels in Texas. Negotiations with the *Carrollton Sun* reached the point of Riddell's penning a sample chapter.[12] But in the end, he seems to have been unsuccessful in working out an arrangement, as there is no further reference to the journal in his papers. Being a man of considerable ego, Riddell would have likely commented on its publication. Subsequently, the journal was probably put aside for future attention. An increasingly busy life, however, prevented his return to it. Upon Riddell's death his papers, or at least the portion containing the diary of his Texas travels, fell into the hands of one of his sons. Sometime after the younger Riddell's death, his widow gave them to Tulane University, where I discovered the diary several years ago while researching the history of southern science.

While travel accounts to early Texas abound, Riddell's previously unknown journal is a noteworthy find. Entering Texas at Galveston, he journeyed west from Houston to San Antonio, and then north-northeast to the Colorado River. The botany of Southeast and Central Texas had been studied earlier by Jean Louis Berlandier, a young Swiss student of nature (1828–29), and Thomas Drummond, the eminent Scottish botanist (1833–34). Riddell, while intelligently commenting on the Texas natural world and turning up an occasional new species, largely corroborated the work of his predecessors. Unlike them, however, he went beyond botany to describe the geology of the area, and did so in a remarkably accurate fashion. The chief significance of his second trip to Texas, as this diary establishes, is that he was the first trained scientist to travel to the area of the Edwards Plateau, antedating by almost a decade the well-known visit of Ferdinand Roemer. This journal represents a preliminary natural history of the region. Like others, Riddell recorded the sights and sounds of Texas during the vibrant and turbulent early days of its independence, but from the seldom seen perspective of the scientist. And Riddell's narrative is high adventure, telling of lost treasure, Indian warfare, colorful personalities, and plotting and scheming.

The diary has its limitations. It abruptly and inexplicably ends on November 12, 1839, some two weeks or so before Riddell completed his stay in Texas. The reader is left hanging. At this time, having failed to locate the legendary San Saba silver mine, Riddell was on the Edwards Plateau in the vicinity of the confluence of the Llano and Colorado rivers. Although he apparently made his way down the Colorado to Austin, his subsequent adventures, route back to Galveston where he embarked for New Orleans, and elapsed time are unknown. Written while on the trail, Riddell's entries are frequently terse. A virtual enthrallment with the flora of early Texas accounts in large measure for this limitation, as Riddell spent much of his leisure time writing descriptions of the plants he encountered.[13] At times, he unquestioningly passes on hearsay as fact. His plant names are a further source of difficulty. Riddell disapproved of the prevailing Linnaean system, preferring instead aspects of those of Amos Eaton, John Lindley, and Augustin Pyramus de Candolle. The result is a patchwork of classifications that frequently made it difficult to arrive at the true identity of a specimen.

I have adopted a conservative approach in editing Riddell's diary. Since the manuscript journal is not readily available, I decided to pass it on as near the original as possible. In particular, I have left paragraphing untouched, although combining frequent one-sentence and short paragraphs would perhaps enhance readability. Punctuation and misspellings, often the result of Riddell's writing phonetically what he heard, have been silently corrected. Some exceptions, such as Brassos (Brazos), Guadelupe (Guadalupe), musquit (mesquite), and mockasin (moccasin) have been left uncorrected for flavor. Explanatory notes have been provided to give the modern equivalency and common names for plants listed and to elucidate the text. Mindful of the vast travel literature on Texas, I have kept descriptive notes to a minimum. Finally, because Roemer's subsequent investigations of the Hill Country have long been recognized as the pioneering work on the area, an effort has been made to correlate those of Riddell with them.

I have received much generous assistance in the preparation of this small volume. Foremost, I wish to acknowledge a large debt to the John Simon Guggenheim Foundation for providing early funding for my forthcoming study of science in the South, of which this piece is a spin-off. I am indebted to Wilbur E. Meneray, head of Archives and Special Collections at Tulane University, for permission to publish Riddell's diary. Barney Lipscomb, collections manager at the Botanical Research Institute of Texas, provided invaluable assistance in the difficult task of identifying Riddell's Texas plants. Albert E. Sanders, curator of natural sciences at the Charleston Museum, earned my great admiration for making sense of one seemingly unintelligible entry in the diary. The staffs of the Southern Methodist and Tulane university libraries professionally and cheerfully responded to my every request for assistance. Kathryn M. Lang, senior editor of the Southern Methodist University Press, graciously read portions of the manuscript and gave me much appreciated advice in placing it. The staff of Texas A&M University Press merits high praise for superior editing and bookmaking. Finally, I thank my wife, Lee, for her unflagging encouragement and valued editorial services.

NOTES

1. Marilyn McAdams Sibley, *Travelers in Texas, 1761–1860* (Austin: University of Texas Press, 1967); Samuel Wood Geiser, *Men of Science in Texas, 1820–1880* (Dallas: Southern Methodist University, 1958–59).

2. Ferdinand Roemer, *Texas, With Particular Reference to German Immigration and the Physical Appearance of the Country.* Translated

From the German by Oswald Mueller (San Antonio: Standard Printing Company, 1935), p. 219; Samuel Wood Geiser, *Naturalists of the Frontier* (Dallas: Southern Methodist University, 1937), pp. 181–214. Geiser, the authority on the pursuit of natural history in early Texas, noted Riddell's earlier trip to southeastern Texas, but was unaware of this subsequent visit to the Hill Country (p. 232). He later learned of the second visit and mentioned it in passing in his *Men of Science*, p. 183. Roemer's place in Texas natural history, however, remains unaltered.

3. Thomas Cary Johnson, Jr., *Scientific Interests in the Old South* (New York: D. Appleton-Century Company, 1936), p. 157.

4. John Leonard Riddell, "Personal Journal," Apr. 18, May 24, 1839, John Leonard Riddell Papers, Manuscripts Division, Tulane University, New Orleans, La.; John Leonard Riddell, "Observations on the Geology of the Trinity Country, Texas, Made During an Excursion There in April and May, 1839," *American Journal of Science and Arts* 37 (1839): 211–17.

5. See C. F. Eckhardt, *The Lost San Saba Mines* (Austin: Texas Monthly Press, 1982); John Warren Hunter, *Rise and Fall of the Mission San Saba. To Which is Appended a Brief History of the Bowie or Almagres Mine* (Bandera, Texas: Frontier Times, 1935); J. Frank Dobie, *Coronado's Children: Tales of Lost Mines and Buried Treasures of the Southwest* (Dallas: Southwest Press, 1930), chap. 1.

6. An entrepreneur, Hopkins came to southeastern Texas from his native New York by way of Alabama. In 1838 he became involved in a scheme to buy Galveston Island and establish a town after discovering that San Luis Pass afforded safe anchorage into Galveston Harbor. The next year he opened a store in Galveston and helped establish a newspaper. The town failed when the channel at San Luis filled up. Subsequently, Hopkins lived for many years in Austin, where he became clerk of the U.S. District Court. He died in 1883. *The Handbook of Texas* (Austin: Texas State Historical Association, 1952), 1:835; "Personal Journal," July 2–3, 4, 1839.

7. "The Late Dr. J. L. Riddell," *Daily Southern Star* (New Orleans), Oct. 8, 1865.

8. Riddell, "Personal Journal," Dec. 22, 1859.

9. John Leonard Riddell to John Torrey, Feb. 9, 1840, in Karlem Riess, *John Leonard Riddell*, Tulane Studies in Geology and Paleontology, no. 13 (New Orleans, 1977), p. 38.

10. "Personal Journal," Feb. 10, 1840. The assumption that nothing came of Riddell's effort to have his Texas plants published by the American Philosophical Society is grounded in a review of the society's publications: *Proceedings of the American Philosophical Society Held at Philadelphia for Promoting Useful Knowledge* (Philadelphia: 1838–present), and *Transactions of the American Philosophical Society Held at Philadelphia for Promoting Useful Knowledge: Cumulative Index* (Philadelphia, 1961).

11. John Torrey and Asa Gray, *A Flora of North America* (facsimile of the 1838–43 edition), Introduction by Joseph Ewan (New York:

Hafner Publishing Company, 1969), 2:189, 232, 246, 256, 271, 369, 382, 421, 444.

12. *Carrollton Sun*, July 19, 20, 24, 25, 26, 1840. Carrollton was the largest of New Orleans' upriver suburbs. A resort and truck farming center, it had a population of 1,470 in 1850. Robert C. Reinders, *End of an Era: New Orleans, 1850–1860* (New Orleans: Pelican Publishing Company, 1964), pp. 7–8.

13. Although his collections from Ohio, Louisiana, and elsewhere have survived, none of the specimens that Riddell collected in Texas are now known to be extant. See I. Hettie Vegter, "Index Herbariorum: Part II (5), Collectors," *Regnum Vegetabile* 109 (1983):760.

# A Long Ride in Texas

# Introduction

John Leonard Riddell was born on February 20, 1807, in Leyden, Massachusetts, the first of ten children. The Riddells, who had been in New England since 1737, traced their ancestry to eighth-century Scottish nobility. By the time John was born the family had fallen on hard times. His father and namesake, who was longer on ambition than accomplishment, was hard-pressed to support his growing family. The elder Riddell was at various times in his life a school teacher, constable, justice of the peace, and farmer. Before John's first birthday, the family moved to a farm in Chenango County in south-central New York. Here the youth spent his childhood in poverty and isolation.[1]

Riddell's education was meager because of limited opportunities and having to work on the family farm, which he hated to the point of being characterized as "a very lazy boy" and receiving "frequent floggings."[2] His earliest instruction was provided by his mother and other relatives. "My early opportunities for acquiring education," he recalled, "were far from superior. Until I was more than 17 years of age I never saw the interior of a more dignified edifice devoted to learning than a common schoolhouse." Riddell was twelve before he learned to write, and seventeen when he began the study of mathematics. He became

3

especially fond of reading, preferring to read on his own over attending formal classes. In fact, Riddell's interest in science dated from 1821 when he read Spafford's general geography on his own, after becoming offended and leaving an uncle's school. This work went beyond geography to include introductions to astronomy, chemistry, and physics. These subjects, he wrote, were "entirely new to me and . . . absorbed my whole attention. I resolved to become a philosopher, and forthwith set my ingenuity on the alert." Obsessed with overcoming his impoverished background, Riddell subsequently worked hard and excelled as a student. By eighteen, he had exhausted local educational oportunities and became a schoolmaster himself. After two years, however, he quit to continue his education at Oxford Academy. The single term Riddell spent there was the extent of his education above the common school.[3]

In 1827, Riddell enrolled at the Rensselaer School (now Rensselaer Polytechnic Institute), the nation's first college of technology. His principal professor was Amos Eaton, a scientist of diverse interests and immense popular appeal. Described by an early biographer of the school as "a naturalist of the old school, almost a 'philosopher' in the Chaucerian sense," Eaton lectured and wrote textbooks on such subjects as botany, zoology, chemistry, and geology. But he is best known for his work in geology and is regarded by some as the leading American geologist of the 1820s. Eaton was Rensselaer's most prominent faculty member. In fact, the school had been founded by Stephen Van Rensselaer at Eaton's suggestion. Eaton was largely in charge of its operation. Under his guidance, the curriculum stressed the relation of science to daily life and was grounded in laboratory instruction and field work. The results were impressive, as a number of the succeeding generation's principal scientists studied there.[4]

In 1829, upon the completion of two four-month terms, Riddell was awarded the Bachelor of Arts degree. Initially, he returned home to open his own school. But after only one year he abandoned the life of a schoolmaster to become an itinerant lecturer in chemistry and botany. This was an era of widespread popular interest in science because of its supposed lessons for personal and national progress. A diverse group of lecturers, ranging from science professors to charlatans, satisfied the public's appetite for science. The best gave lectures by invitation; most, however, went wherever they could gather a paying audience. Riddell was motivated to become a lecturer in science for a variety of reasons. He did not like teaching, exclaiming on one occasion: "O Heavens, for my existence sake, deliver me from the thankless task of teaching a room full of hard-headed, thick-pated, mischief-making boys." Lecturing offered a potentially profitable alternative. Moreover, it was a way to continue the pursuit of science.[5]

Riddell delivered his first series of lectures at Ogdensburg, New York, in the fall of 1830. Although there was little original in them, consisting largely of basic scientific principles from Eaton's lectures and his own readings, they were favorably received. As was common in such ventures, Riddell's success was in large part the result of showmanship—in his case the skillful use of spectacular experiments and scientific apparatus as crowd pleasers. Riddell was encouraged, even though barely eking out a living, and ranged farther afield. Between 1830 and 1832, his intellectual wanderings took him first through upstate New York and the border towns of Ontario, Canada, and then to Erie and Pittsburgh, Pennsylvania; Wheeling, Virginia (now West Virginia); and Marietta, Ohio. His usual practice was to solicit subscriptions from prominent citizens in a community and deliver a public lecture before commencing a subscription lecture series. Should the affair go well, he would then ask these citizens for letters of introduction to their friends in a neighboring town. Somehow, perhaps because he conducted himself in public with an off-putting "hauteur and reserve," Riddell "acquired the name of being eccentrick," which sometimes hampered his efforts. But even worse, he lost favor in several communities because of an unfortunate habit of philandering. Strikingly handsome and naturally charming, he was immensely appealing to young women. He had his first of a long line of affairs at age fifteen. Even marriage did not curb his womanizing.[6]

By the time he arrived in Marietta in June, 1832, Riddell was tiring of his nomadic life and failure to advance financially. "I have," he had lamented a year earlier, "formed plans of becoming eminent; I have built several castles in the empty air. I have coveted solitude, and devoted my attention to the sciences, cultivated by men whose names grace the annals of history. But I find I have been founding my hopes upon the fickle wine. Untoward circumstances, yes poverty, unrelenting poverty, now opposes an insurmountable barrier to my advancement."

The desire for new opportunities "to redeem an empty name from that bottomless gulf of oblivion in whose dark waters is overwhelmed the memory of the mass of human beings" increasingly dominated his thoughts.[7] To secure living expenses in Marietta, Riddell accepted a position as temporary professor of astronomy, geology, botany, and chemistry at a local female seminary. But he dedicated the bulk of his energies to the collection and sale of sets of plants indigenous to Ohio.

This was a fairly common practice because of nineteenth-century America's great interest in the natural world. The study of natural history, as the natural sciences were called at this time, was a way to combat the loneliness of rural life, a means of observing God's handiwork and beneficence, and a possible source of personal gain. Riddell's personal fascination with nature dated from his student days at Rensselaer. He exclaimed on one occasion: "I wish to view the scenes of nature that furnish the imaginations with grandeur and sublimity."[8] In 1830 Riddell began his herbarium, or collection of botanical specimens. In this age of collection and classification, the most prized possession of the student of nature was his cabinet. The

quality of these collections of natural history specimens
varied greatly, depending upon the ability and industry
of the owner. Some were general in character, consisting of
flora, fauna, minerals and rocks, shells, fossils, and Indian
relics. Others were specialized—botany in Riddell's case,
for instance. His was a source of immense personal pride
because of its numerous and varied plant specimens. Many
of these were collected during botanizing expeditions, or
field trips, which had long occupied his leisure time on
the road. Armed with a handful of botany manuals, maps,
and botanical supplies—a press for preserving specimens,
botanical dissecting instruments, watercolors, and drawing
paper—he scoured the countryside for specimens. Riddell
was especially skilled at preparing dried plants. He took
great pride in the quality of his specimens and constantly
sought ways to improve upon it, experimenting with new
methods and equipment.[9]

Riddell's labors in Ohio natural history led to his in-
augural publication, a survey of local botany, which appeared in three installments in the *Marietta Republican*.[10]
It was the first of nearly eighty publications in a long and
active life in science. But his hopes of reaping a profit from
the study of nature amounted to little. And at the end of
his temporary professorship in the fall of 1832, he accepted a full-time one in chemistry and botany in the
Ohio Reformed Medical College in Worthington.[11] Established in December, 1830, this was one of the growing
number of sectarian medical schools that sprang up in
protest of the heroic practices of the allopaths—puking,
purging, bleeding, and large doses of potentially dangerous
drugs—which met with little success in the day-to-day
struggle against common complaints and failed miser-
ably when confronted by yellow fever, cholera, and typhoid
fever—the great killer epidemics of nineteenth-century
America. The Worthington school denounced "metallic
minerals [and] the lancet or the knife," the chief symbols
of an allopathic practice, as "dangerous and inefficient." In

their place, it offered a botanic regimen, claiming "vegetable substances alone, are void of danger, when properly administered."[12]

The position appealed to Riddell for two reasons: it promised him a steady income and time to continue his research and writing. Upon arrival, however, he found the school to be a greater source of concern than of promise. Like many of the nation's myriad medical schools, it was small and of low quality. Students usually numbered between fifteen and twenty and were mostly middle-aged men. The faculty was equally unpromising. Riddell wrote a friend during his early days at Worthington: "There are but two professors here, and I can assure you, I was happily disappointed in respect to their qualifications." Whether I remain here during the spring or not," he added matter-of-factly, "will depend entirely upon the dollars and cents which may be offered."[13] But he expected his stay to be short, writing in his journal within days of his arrival: "I still retain too much respect for the accumulated experience of men of science to allow the idea of connecting myself permanently with this institution."[14]

Riddell remained at the Ohio Reformed Medical College for two years, until the spring of 1834. During this time he fretted constantly over professional advancement and financial gain. The death of his father in May, 1833, and the greater responsibility of helping his mother and younger brothers, added to Riddell's financial worries. Midway through his stay in Worthington, he lamented: "I have scarcely money enough to pay letter postage and with a meagre prospect for more; connected with an institution which is far from meeting the requirements of my ambition."[15]

Life at Worthington soon settled into a deadening routine. "My time," Riddell wrote in his personal journal less than a year after he arrived, "goes almost monotonously.—No singular events.—Nothing new: It is read,

eat, and go to the college; read, eat, and go to the college."[16]
Riddell sought to combat the gloom of his surroundings
with a redoubled quest for professional and personal advancement. He investigated positions at other institutions
and considered returning to lecturing. He even enrolled in
the medical lectures of his two colleagues, and toyed with
the idea of a career in medicine.[17] But Riddell's greatest
hope lay in the continued pursuit of his scientific interests.
In fact, it is during these years that his insatiable appetite
for the study of science became evident, as he investigated
astronomy, ballooning, botany, chemistry, epidemiology,
meteorology, physics, and physiology. He also began to at-
tract attention in scientific circles. A report and analysis of
a meteor shower for a local paper was cited by Denison
Olmsted, the highly regarded Yale astronomer and geologist, in his account of the same phenomena.[18]

As before, Riddell spent his leisure time botanizing.
His field research fulfilled several purposes. First, he
wished to satisfy his own growing interest in botany. He
prepared a catalog of local plants, which he later published, and made long-range plans to prepare a flora of the
West. Second, he needed specimens to revive his scheme
to sell sets of Ohio plants.[19] Third, and a common practice, he served as a field collector for a number of prominent natural historians who needed specimens for their
work. His botanical correspondents included Constantine
S. Rafinesque, Transylvania University's brilliant but
eccentric former professor of natural history; Charles
Wilkins Short, the most important collector of indigenous plants of the early West; Samuel Hildreth, a well-
known Ohio historian-naturalist; and Robert Peter, the
newly appointed professor of chemistry at Transylvania
University. Of special significance, he collected for John
Torrey, a national leader in the study of American botany,
who in return helped him in the identification of specimens. Riddell's abilities as a collector were permanently

recognized when one of his botanizing partners, a visiting German botanist, named a variety of goldenrod *Solidago Riddellii* in his honor.[20]

On March 31, 1834, Riddell was awarded the M.D. degree by the Ohio Reformed Medical College.[21] Days earlier, he had resigned his professorship. Riddell had been increasingly certain since the beginning of the year that he would leave Worthington at the conclusion of his course of lectures in March. A combination of ambition, money, and the school's low image as a sectarian diploma mill made leaving likely; the proprietors' failure to make him an acceptable financial offer made it inevitable.[22] He later wrote: "I might have remained at Worthington; but I thought it hardly field enough for me, and moreover, I did not like to forego the friendship of the eminent men in the medical profession."[23] Riddell's decision to leave was a major disappointment to the students. He had been the school's most popular teacher, and they had petitioned the faculty to take the necessary steps to keep him. "I have obtained a degree of popularity and respect among the students," he proudly recorded in his personal journal, "which I never dared to anticipate."[24]

Riddell was not sure where he would go, asserting: "Heaven only knows my destination." Of one thing, however, he was sure: "[I]t will result ultimately to my advantage."[25] After much pondering, he chose Cincinnati. This choice seemingly stemmed from two factors, both of which augured well for his future: Cincinnati was the largest and most important city in Ohio; it was also the home of Daniel Drake, probably the most accomplished scientific figure in the trans-Appalachian West.[26] Having learned that it did little good to seek patronage where he was unknown, Riddell had begun the previous fall to cultivate Drake's friendship as part of a personal campaign "to become more generally and favorably known." He asked Hildreth to speak with Drake on his behalf and sent him one of his publications and a suit of Ohio plants. Drake responded with a warm note of appreciation and extended an offer to write for the *Western Journal of the Medical and Physical Science*, the region's principal scientific organ which he had founded and edited.[27] Riddell immediately submitted an essay on collecting and preserving plants from a proposed study of practical botany. Drake was favorably impressed with the piece and published it. It was the first of several to appear in this highly regarded scientific journal.[28]

Riddell arrived in Cincinnati on April 6, laden with nine trunks and boxes containing his herbarium, apparatus, minerals, and chemicals. "I have once more cast myself on the tide of chance," he exclaimed. As soon as he had found a place to live, Riddell called on Drake—armed with letters of introduction.[29] Initially, the eminent scientist greeted him warmly, but soon Riddell's transparent ambition and reputation as a womanizer apparently alienated him, as relations between the two cooled for a time. Undaunted, Riddell settled into a routine of making acquaintances, reading, botanizing, and obtaining subscriptions for a lecture series on botany.[30]

The unpromising early results of these popular lectures, launched in May, convinced him that if he hoped to remain in Cincinnati, he would have to return to selling plant specimens to get by. Riddell was embarrassed at this turn of events, bemoaning: "O I do hate to procure subscriptions for collecting plants. I hate it cordially, because I know it must lower me in the estimation of some and prevent my rising in the estimation of many whom I hope to become acquainted with hereafter. But I have to choose among evils. Money I must have: so I sally forth in fine days to *peddle* collections of plants."[31]

The continued financial distress of his mother, who was reportedly almost destitute, added to his financial woes. Riddell oscillated between despair and determination, remarking at the end of three months in Cincinnati: "O! I feel heart sick. Pride, inclination, duty, necessity, hope and fear, by turns predominate in my breast and I feel at times as though I should sink beneath their weight. Something I must do. Something I will do, if my health and strength are spared."[32]

This "something" turned out to be a combination of endeavors. Riddell enrolled at the Medical College of Ohio, founded by Drake in 1819, in the renewed belief that the practice of medicine might be a solution to his financial problems. Although a recent graduate of the Ohio Reformed Medical College, he thought it best to start over because of the latter school's sectarian character and low image. In fact, upon arriving in Cincinnati, he renounced his earlier degree. "Drake," he wrote in his personal journal, "asked me if I was a physician. I told him not."[33] Riddell was also named an adjunct in chemistry and lecturer in botany in the Medical Department of the Cincinnati College, a rival to the Medical College of Ohio a disgruntled Drake established in 1835.[34] His popular lectures, teaching duties, and attending medical lectures occupied the bulk of Riddell's time. The little that remained was divided between botanizing and the pursuit of his other numerous interests in science. He undertook several lengthy and wide-ranging "botanical peregrinations" through Ohio. These were to collect specimens to sell and to increase his knowledge of local botany.[35] Riddell also completed a catalog of western plants which he had begun while at Worthington. Drake agreed to publish it. The resultant "Synopsis of the Flora of the Western States," containing listings of 690 genera and 1,800 species (13 of which were new), was the first such undertaking devoted exclusively to the flora of the West. Although drawing heavily on the work of Charles Wilkins Short, Thomas Nuttall, and other naturalists, it was nevertheless considered a contribution of major importance and was well received. A year later, a second catalog, which contained seven new species, extended this work.[36] Riddell also revived a long-standing interest in scientific experimentation and theoretical speculation, asserting: "I am now in my old employment of building hypotheses, and devising experiments and apparatus."[37]

Riddell's labors this time proved profitable, and seemingly quite suddenly. In the fall of 1835, he complained: "My coat is so poor I am almost ashamed to wear it. . . ." But by the end of January, 1836, he exulted: "I am . . . out of debt, and have money enough to meet my contingent and unavoidable expenses probably for some time." "Neither am I particularly plagued with matters pertaining to love," he cheerfully added.[38] In addition, Riddell's professional standing was on the rise. He was increasingly accepted in Cincinnati scientific circles, having been elected a junior member of the Medical Society of Cincinnati and a member of the Western Academy of Natural Sciences. Perhaps most important, he was intimately involved in the campaign to persuade the Ohio legislature to establish a state geological survey.[39] From their appearance in the 1820s, geological surveys increasingly captured the antebellum popular mind. These were the state complements of the national exploring expeditions in the West and represented the chief form of public aid to scientific investigation. In general, they were established to survey natural resources, to stimulate economic activity, and to aid the agricultural class through soil analyses and providing information on new crops. Since the geological surveys were generally conducted by trained scientists, however, much valuable research in geology, mineralogy, and paleontology was conducted under the guise of applied science.[40] Riddell hoped to be named state geologist. This position would allow him to turn his love of natural history into a full-time vocation, and its substantial salary would ensure continued financial stability. He had Drake's support and interpreted as a good omen his appointment in 1836 to a geological survey planning committee. Subsequently, Riddell devoted much time to a geological reconnaissance of his assigned area of the state. His hopes for an appointment as state geologist were to be dashed, however, because the legislature refused to create a geological survey.[41]

Riddell was graduated from the Medical College of Ohio on March 5, 1836. The subject of his required thesis for the M.D. degree was an assessment of the rival theories of miasmata and contagion in the appearance of disease. The former, long-standing and dominant, attributed disease to noxious odors arising from decaying animal and vegetable matter. The latter, relatively new and revolutionary, stressed the role of organisms in the etiology of morbid conditions. Riddell's support of the animalcular theory of contagious infections, which dated from his days at Worthington, was largely circumstantial, but it put him well ahead of most of his contemporaries in disease origin. "I am," he wrote as early as the end of 1833, "inclined to favor the hypothesis, that contagious and infectious diseases are caused by animalculae."[42] Drake, who had earlier suggested an animalcular cause for cholera, lent his support to Riddell's argument before a doubting audience at the Cincinnati Medical Society and published the thesis.[43] Riddell, ever desirous of "the gratification of possessing a solid reputation," noted with considerable pleasure that a leading New York medical journal had reprinted the piece, longingly adding: "If it has the fortune of being republished across the Atlantic, I shall be satisfied."[44]

One of Riddell's first acts after taking his degree was to marry. After a string of affairs, most of which had been short-lived and unsatisfying, Riddell looked forward to marriage. "In all my amorous adventures," he exclaimed on one especially painful occasion, "I seem to be wonderfully unfortunate."[45] The ideal in a wife to him was someone who would be a loving companion and help him advance financially. In Cincinnati he met Mary Bone, a student at a local boarding school. Mary was an attractive and eager-to-please orphan from New Orleans with a claim to a sizable estate. After a six-month romance that was carried on in secret because of the headmaster's dislike for him, Riddell eloped to Kentucky with his "Creole Mary" on June 20, 1836. The couple honeymooned throughout the summer while Riddell undertook the botanical and geological exploration of northeastern Ohio for the proposed geological survey of the state.[46]

The newlyweds returned to Cincinnati in August, 1836, only to find themselves outcasts, shunned by the wives and daughters of the professors of the medical school because of rumors about their courtship and marriage. (Mary had come to Riddell's room before they were married.) The couple denied any wrongdoing and resolved not to be affected by the harsh and hurtful behavior toward them.[47] Within a matter of days a way out of this sticky situation presented itself when Riddell was offered the chair of chemistry in New Orleans' Medical Department of the University of Louisiana. Founded less than two years earlier, this was the first medical school in Louisiana. In 1884, it became the Tulane University School of Medicine.[48] Riddell considered his prospects in Cincinnati "perhaps good enough." A supplementary catalog of Ohio plants that appeared in Drake's journal had been well received. Moreover, he believed that the state legislature would create a geological survey during the next session and that he had "a finger in the pie" for the position of state geologist. Still, as he put it: "I have no inseparable objections to try New Orleans a winter or so."[49] In reality, the offer was appealing for three reasons. First, he was chafing under the embarrassment resulting from the treatment of him and his wife. Second, Mary's inheritance was tied up in court, and Riddell, who was determined to have it, recognized that a Louisiana residence was almost necessary in pursuing her claim. Finally, he saw the Crescent City as professionally promising, observing in his letter of acceptance "by zealous . . . endeavors, New Orleans may ere long become the great Emporium of medical instruction for the South."[50]

Riddell's "winter or so" in New Orleans lasted nearly thirty years—the remainder of his life. At first, the intellectual isolation of the Gulf Coast South and its obstacles to the pursuit of science bothered him and he thought of leaving. "Though my prospects in a pecuniary point of view," he wrote Drake in June, 1838, "are fair enough; yet the want of associates in the prosecution of the natural sciences causes me to think often of Cincinnati."[51] A few months later, he lamented to Amos Eaton, his mentor: "I publish nothing now-a-days, not because I am idle, but because it costs nearly three times as much money here as at Philadelphia to procure printing."[52] For the moment the legal problems surrounding his wife's estate kept Riddell in New Orleans.[53] By the time the affair had been resolved— and in his wife's favor—he had adjusted to the circumstances of life in that city.

Riddell matured as a scientist in New Orleans and came to have a profound influence on scientific thinking there and in the Deep South. While his professorship was demanding, intellectual curiosity and ambition resulted in myriad scientific undertakings. The personal journal that he kept for nearly two decades after graduating from Rensselaer is filled with sketches, descriptions, and speculations pertaining to an astounding variety of subjects. At the outset, Riddell resumed the study of natural history. He wrote in his personal journal in the summer of 1840: "The study of nature is what I wish to excel in. Let me but carry out one half of the plans I have formed; let me accomplish one half of the research I have proposed to myself, and I shall then feel more satisfied."[54] Riddell returned to botanizing and resumed his extensive botanical correspondence. It was local, regional, national, and eventually international in composition. Prominent among his correspondents were John Torrey and Asa Gray, leaders in the study of the American natural world. A number of the Louisiana specimens he sent them were included in their ambitious *Flora of North America*.[55] Riddell's personal goal was to prepare a catalogue of the plants of Louisiana as he had for Ohio earlier. "The botanical novelties constantly presenting themselves here," he wrote Drake shortly after settling in New Orleans, "have occupied my attention."[56] In 1851, he sent a manuscript of state flora to the Smithsonian Institution. It described some 2,300 species, a number of which were new species and varieties he had discovered. But apparently because it was not entirely original, the catalog was never published. The following year, Riddell published the original portions of the manuscript at his own expense.[57] As before, he also spent much time trying to perfect his techniques for preserving plants. In addition, he experimented with a new lithographic process for the printing of specimen plates.[58]

Riddell's professional and personal interests carried him to Texas during much of the spring and fall of 1839. Upon returning home from his second trip in November, he learned that President Martin Van Buren had appointed him melter and refiner for the New Orleans branch of the United States Mint. This position was apparently a reward for his heavy involvement in local Democratic politics. Riddell's tenure, while marred, and even threatened, by technical problems growing out of his inexperience in metallurgy, several heated confrontations with subordinates, and politics, further contributed to his financial independence. At the same time, it severely curtailed his scientific activities. "I am so annoyed by the responsibilities of my situation as melter and refiner," he wrote in his personal journal a little over a year after assuming his duties, "that I find little leisure for reading or writing; and am unable to make much progress in any of my projects."[59]

Not until he was removed from the post in 1848 did Riddell return to his study of botany. By this time, other scientific interests were monopolizing his time. Still, Riddell's accomplishments in botany did not go unrecognized. He remained a valued botanical correspondent of Gray and Torrey. During a trip to the North that Riddell made in the summer of 1840 on mint business, Samuel G. Morton, the pioneer American ethnologist, honored him at a party in Philadelphia, which was attended by a number of the nation's leading scientists. Among those in attendance were R. M. Patterson, the scientist-director of the U.S. Mint; Alexander Dallas Bache, the geophysicist first president of Philadelphia's Girard College and future superintendent of the United States Coast and Geodetic Survey; Robley Dunglison, eminent author and teacher at Jefferson Medical College, who had been a member of the first faculty of the University of Virginia; and Thomas Nuttall, the celebrated English botanist known for his pioneer work in collecting the plants and seeds of the American West. Later, Nuttall, a longtime correspondent, assigned the name *Riddellia* to a western composite genus in Riddell's honor.[60]

Riddell's activities in natural history went beyond botany. He had long been interested in geology. In New Orleans, as in Cincinnati earlier, Riddell played a leading role in the campaign to persuade the state legislature to establish a geological survey. His involvement dated from March, 1838, when he drafted a bill creating a state survey. It intensified the following winter, as he mounted a personal campaign that included memorializing the legislature and writing promotional pieces for the local press, to muster public support for the proposed legislation. "I am trying," he wrote in February, 1839, "to get up a geological survey of the state." Again, Riddell hoped to be appointed to the position of state geologist.[61] Finally, in 1841, a geological committee was appointed, and he was named chairman. Despite this promising step and strong popular pressure for a state survey, the movement foundered in the legislature. In 1855, in conjunction with the New Orleans Academy of Sciences, Riddell helped spearhead a renewed campaign for a state survey. But the fiscally conservative legislature again turned a deaf ear. In the South, only Louisiana and Florida failed to conduct geological surveys before the Civil War.[62]

But for all his fondness for natural history, after his arrival in New Orleans, Riddell increasingly gravitated toward natural philosophy and medicine. In natural philosopy,

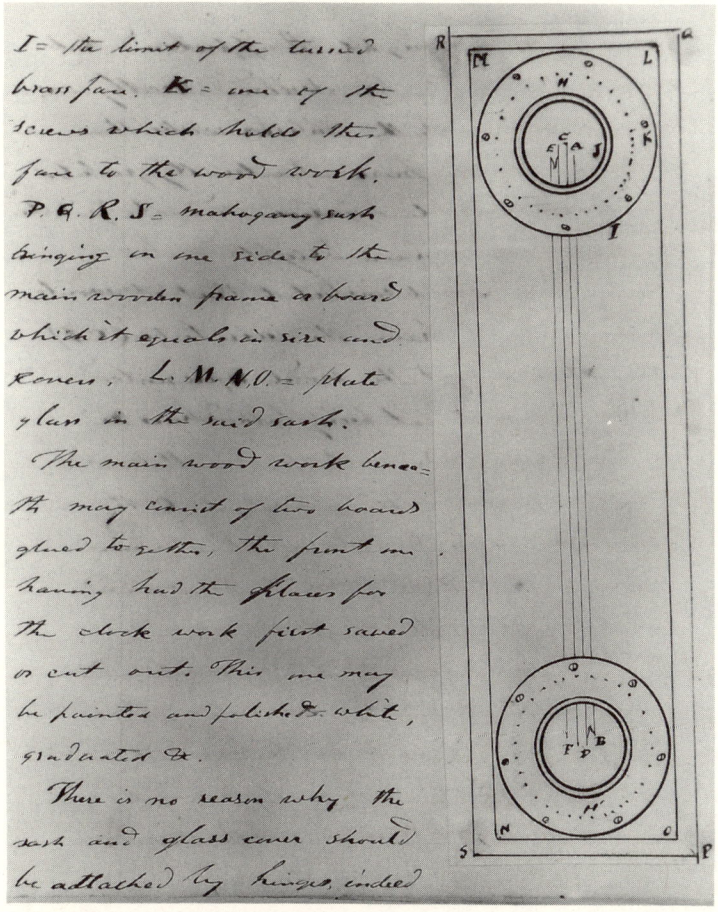

Drawing of an experimental spirit clock from Riddell's personal journal. Courtesy Special Collections, Tulane University

he was largely attracted to physics, in both its applied and theoretical forms. In applied physics, he was an active inventor, and his personal journal contains frequent references to inventions. Among Riddell's creations were a plant dryer, a pocket compass, a fishrod balance, a spirit clock, a phosphorescent taper, a prototype of the typewriter, an apparatus for measuring the specific gravity of coins, and a "galvanic device" for dissolving bladder stones.[63] But he is best known for the invention of the binocular microscope. Riddell's interest in improving on the single-eyepiece microscope dated from about 1850 when he conducted a series of microscopic examinations of New Orleans waters. By 1851, he had designed the binocular microscope. "In principle," Karlem Riess, his physicist-biographer writes, "it divided light from a single objective by means of a combination of four glass prisms, passing the light beams to the eyes through two parallel tubes, each having its own ocular."[64] On October 2, 1852, Riddell unveiled his invention before the New Orleans Physico-Medical Society. The next summer he reported on it to the annual meeting of the American Association for the Advancement of Science. Sadly, design problems that inhibited reliable stereoscopic vision prevented any significant use or mass production—a bitter blow to his ego. Even worse, because of this situation, credit for the discovery of the principle of the binocular microscope has been given to others.[65] But according to J. J. Woodward, a pioneer American microscopist who is remembered for his path breaking work in photomicrography, this was a major oversight. Writing in the 1880s of Riddell's work, Woodward remarked: "He undoubtedly deserves the credit of having discovered and first published the optical principle, on which all the most successful binoculars made prior to the present year depend."[66]

Riddell's philosophical speculations extended across a broad front.[67] But his theoretical bent is perhaps best seen in his fascination with aerial navigation, the theory of gravitation, and speculations on the constitution of

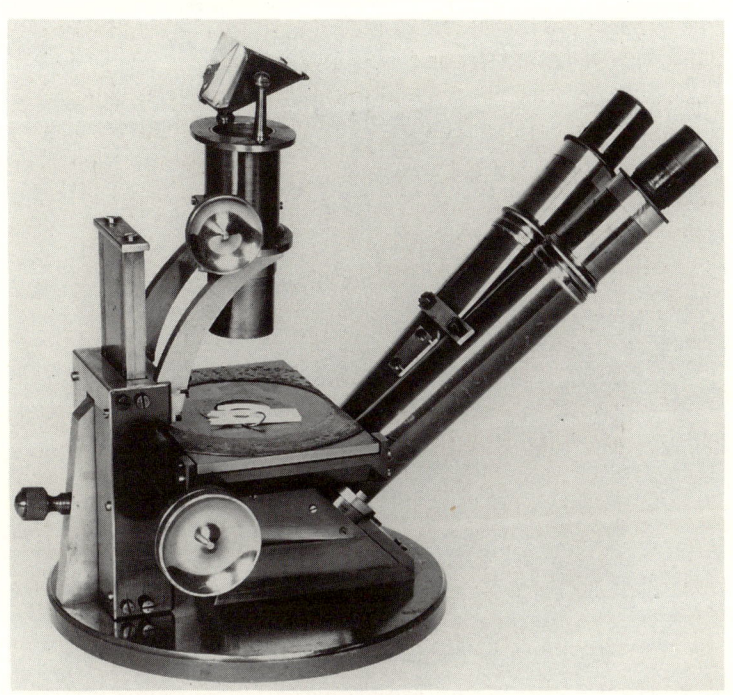

Riddell's binocular microscope. Courtesy Historical Collections, National Museum of Health and Medicine, Armed Forces Institute of Pathology

matter. The prospects of traveling to the moon in an "aeronautic car" had long intrigued Riddell. As early as 1831, while in Ogdensburg, New York, he was recording rambling musings on aerial navigation in his personal journal and planned to prepare a popular lecture on it. Throughout his subsequent intellectual odyssey, Riddell never lost interest in space travel. He continued to speculate on the mechanics of it, and spaceship design most especially. In the fall of 1843, he became embroiled in a heated controversy, waged for over two months in the New Orleans press, with a rival extraterrestrial enthusiast.[68] At last, in April, 1847, his proposed lecture became

New Orleans, Friday 24th May, 1889.

I arrived from Texas, per. Steam ship Columbia, on Tuesday evening 14th inst.

I have been thinking on aerial navigation

Model above.

The balloon made of thin gilded sheet copper.

C. a copper lined room for the aeronaut. P. the bow or head.

v v. rudder for elevating or depressing

L L. rudder for steering laterally

V = vertical section of copper balloon.

The sides may collapse or expand laterally, so as to allow of the

Sketch of a space ship from Riddell's personal journal. Courtesy Special Collections, Tulane University reality when he presented to the People's Lyceum in New Orleans "Orrin Lindsay's Plan for Aerial Navigation With a Narrative of His Explorations in the Higher Regions of the Atmosphere, and His Wonderful Voyage Round the Moon." A fictional Ohio student, patterned on Riddell, Lindsay experimented with various spaceship designs and sent a dog aloft in an experimental model before settling on a globular metal balloon ten feet in diameter that was said to be unaffected by gravitation. The account of the voyage is strikingly modern in discussions of oxygen supply, the temperature of the rarefied atmosphere, the possibility of collisions with meteorites, scientific calculations and observations, and the physiological effects of altitude. Lindsay's description of the surface of the moon and view of the earth from the moon are especially well done.[69] Subsequently published, this is considered Riddell's cleverest piece of writing. "It is," J. D. B. De Bow, the publisher of one of the South's most successful popular magazines, remarked, "an admirable production, and without question the most capital 'moon story' we ever heard of."[70]

While Riddell was in earnest about the possibility of space travel, the subject also provided him with a vehicle for an assault on Newton's theory of gravitation. One reviewer remarked of his essay: "It is ingeniously written, and we suppose may be regarded as a scientific burlesque upon the old theory of gravitation which Dr. R. is endeavoring to demolish."[71] Indeed, Riddell's studies and speculations in mathematics and mechanics had convinced him that Newton was wrong, and he proposed his own radiant cause of gravitation. Underlying this theory was the conviction that gravitation was not material. He vehemently denied any notion of inherent attraction, or a propensity in bodies to approach each other, as the cause of gravitation. "It is utterly absurd," he exclaimed, "and I therefore deny it."[72] Riddell postulated that gravity was caused by "the radiations of ethereal particles in all directions; having the power to impart their momentum to ponderable matter." "A cannon ball or other object in the air," he elaborated, "has a tendency to approach the earth because it receives the impulses of those moving particles from all directions except toward the earth, where the momentum of those due from that direction has already been intercepted."[73]

Riddell's thoughts on gravity were followed by a symbolic formulation of matter. He held that matter was composed of atoms; that each atom was composed of an infinity of particles; and that these were likewise composed of an infinite number of subparticles.[74] While the paper, Riess asserted, "is tedious and on occasion faulty in its reasoning and general logic," Riddell's concept of the atom "is indeed unique because it is our present-day picture predicted many years before the discovery of our atomic particles."[75] Riddell's materialism left little room for a supreme being. "It seems to me," he mused in his personal journal, "that all nature is constituted on the model of a republic, rather than on the model of a monarchy. In other words, I do not think there is in, or out of nature, a separate and independent being, called Deity; but that Deity is the infinite total of all: —that, I, for instance, as well as the meanest or most exalted thing known, am an infinitesimal part of Deity."[76]

In general, Riddell's philosophical speculations were largely just that and were generally dismissed or ignored. Supersensitive to criticism, he vigorously and unrelentingly defended his beliefs. In one instance, this obstinateness precipitated perhaps the bitterest controversy of his career. It began when Dr. Richard Welles Ely, a prominent New Orleans physician and gifted scientist, criticized Riddell's views on gravitation and matter in the local press. Ely ridiculed them as so much nonsense. Riddell responded with outrage. The ensuing controversy, consisting of lengthy arguments and detailed rebuttals, dragged on for weeks until the editors grew tired of it and called a halt.[77] This unfortunate incident pointed up Riddell's darker side. For all his skill and accomplishment as a scientist, he was, in Riess's words, "eccentric, always outspoken and pompous, extremely cynical, and eager for personal improvement."[78] Riddell's self-promotion and "prickly bear" personality even carried over into his relations with his colleagues in the university's Medical Department, with whom he was constantly squabbling. Things got so bad in 1851 that the faculty demanded his resignation. Only the intervention of the Dean averted a showdown.[79]

In striking contrast to his philosophical speculations, Riddell's work in medicine was among his most enduring scientific contributions. He is best known here for his investigation of miasma and contagion and microscopic studies. Riddell, it will be recalled, became interested in disease causation during his stay at Worthington and published circumstantial evidence in support of the animalcular theory after moving to Cincinnati. By the time he arrived in New Orleans, he was thoroughly committed to the belief that contagious and infectious diseases were organic in origin. His subsequent microscopic examinations of local waters, the blood and black vomit of yellow fever, samples of sputa and feces from cholera and tuberculosis patients, and specimens from hospital cases removed any doubts.[80]

A deadly outbreak of yellow fever afforded Riddell an opportunity to press his etiological theory. In 1853, New Orleans was victimized by one of the most lethal yellow fever epidemics in its history. In the wake of this deadly visitation, a sanitary commission, of which Riddell was a member, was appointed by the mayor to determine the cause of yellow fever and to prescribe preventive measures. While the commission's report embraced the prevailing miasmatic explanation, Riddell refused to retreat from the belief that the disease was animalcular in origin. In a dissenting statement he reiterated that the "poisonous matter" responsible for the epidemic was "doubtless some species of living organism," postulated that the "germs" of the outbreak "have probably been derived from countries further south," and recommended "some kind of quarantine in certain months of the year" and a general cleansing of the city "by efficient drainage, and sanitary regulations carried into effect" to prevent a recurrence.[81]

Several months later, at a meeting of the New Orleans Academy of Sciences, he subjected Edward H. Barton, the head of the Sanitary Commission and his colleague in the university's Medical Department, who was perhaps the most prominent spokesman for the role of miasmata in the appearance of disease, to a withering barrage for his contention that the recent yellow fever epidemic "was not imported but indigenous; that it was generated here by the enormous quantities of filth, and the extraordinary disturbance of the earth, on the one hand; and by hygrometric saturation, with extraordinary heat, on the other hand." Riddell challenged Barton to explain how "mere harmless heat and moisture should, at one time, throughout a whole season, produce Yellow Fever; at another Plague; and at another Cholera." He countered that an organic explanation was far more plausible, asserting: "That organisms do exist in the atmosphere in quantities and conditions such as may account for the production of Epidemics, is not merely a deduction or reason; it is proved by chemical tests."[82] Such views were not widely accepted in antebellum America, putting Riddell in the company of a handful of farseeing medical researchers who anticipated the impending bacteriological revolution that was to remake medicine.[83]

Riddell was also a vigorous advocate of preventive medicine. New Orleans, his adopted city, had long been synonymous with disease and death. In the antebellum era it was widely regarded as "the graveyard of the Southwest." According to Barton, also a leader in the fight for public health reform, New Orleans was "one of the dirtiest, and with other conjoint causes, . . . consequently the sickliest city in the Union."[84] Riddell pursued a variety of preventive medicine activities to improve health conditions. His service on the Sanitary Commission has already been noted. Like most sanitarians of the era, Riddell was convinced of the correlation between filth and disease and worked to see that the city was kept cleaner. In 1854, he prepared a technical report for the Sanitary Commission on sewerage disposal in which he included a detailed plan for an improved sewerage system.[85] Efficient drainage and flood control further concerned Riddell. In 1844 and 1855, he was appointed to state committees to devise means to protect New Orleans against the inundations of the Mississippi.[86] The city's water supply—largely cisterns and wells—also commanded his attention. If kept clean, cisterns posed no significant health threat. Surface drainage, however, fouled most of the wells with a wide array of pollutants. Riddell experimented with ways of filtering well water and was one of the prime movers in the abortive efforts of the New Orleans Academy of Sciences to drill an artesian well in the heart of the city's business district.[87] Medical reform was a final way in which Riddell sought to advance public health. Specifically, he argued for a reliance on medical botany in this age of heroic doses of calomel, a mercury derivative. This seeming panacea of the medical profession was administered to the point of salivation, the first symptom of mercury poisoning. In his introductory lecture to the Medical Department in 1846, he urged the students to become familiar with the healing powers of nature's bounty and to start their own herbaria.[88]

The era of the Old South coincided with the early efforts to professionalize science in America. Scientific societies were crucial to this movement. Riddell was perhaps New Orleans' most active promoter and supporter of scientific organizations. He was a charter member of the American Association for the Advancement of Science. Founded in 1848, this was the first national scientific association. A year later, he helped establish the original Louisiana State Medical Society and was named chairman of its committee "on the indigenous botany of the state and its materia medica." He was also active in the affairs of the New Orleans Physico-Medical Society, started in 1819 for

American physicians in the city. Finally, Riddell was one of the founding members of the New Orleans Academy of Sciences. Organized in 1853, this was one of two major scientific societies in the Old South. (The other was Charleston's Elliott Society of Natural History.) He took an active part in the affairs of the organization and was chosen in 1853 to represent it at the Cleveland meeting of the American Association for the Advancement of Science, where he read four papers. As its minutes show, Riddell became the mainstay of the academy. He was elected president in 1855 and was reelected annually until his death in 1865. "During his years as president of the Academy," Riess writes, "Riddell made the community conscious of the need for such an organization, and laid a firm foundation for its growth and development."[89]

In August, 1860, President James Buchanan named Riddell postmaster of New Orleans. The appointment, like his earlier one as melter and refiner for the New Orleans Branch Mint, was probably a reward for Riddell's lifelong involvement in Democratic politics. Although a Unionist who considered secession "a blunder," he held this post until New Orleans was occupied by the northern forces in April, 1862. Following the occupation, Riddell became a vocal Union sympathizer. In a sham election in November, 1863, he was elected governor of Louisiana. The Lincoln government, however, repudiated the proceedings and nullified the returns. Riddell's "traitorous" behavior infuriated much of the community. This animosity turned to fury in October, 1865, at the meeting of the Louisiana State Democratic Convention. Held in New Orleans, it was chaired by Riddell. In his opening speech, he castigated Louisiana for withdrawing from the Union. The audience turned on him. Riddell fled to the office of the *Daily Southern Star* to prepare a statement. In the process of drafting it, he died of heart failure. He was fifty-eight. Riddell had been in failing health for some time and the excitement of the moment proved too much for him.[90]

Although little known today, Riddell is considered by many to be one of America's foremost scientists at the time of his death.[91] This was even the case in New Orleans, where despite residual bitterness, he was praised for his scientific accomplishments. Labeling Riddell "no ordinary man," one obituary eulogized: "His intellect was searching, analytic and vigorous, and his scientific attainments were of a high order."[92] A more fitting testimonial to Riddell was penned by Edward H. Barton, his colleague at the University of Louisiana, co-worker in public health reform, and adversary in the controversy over the origin of yellow fever. Years earlier on the occasion of one of his scrapes at the Branch Mint, Barton had remarked: "With respect to Dr. Riddell's scientific qualification he has not his superior in this community; and if he cannot always make himself as agreeable as others, it is because he has sacrificed the specious to the solid."[93] But perhaps Karlem Riess, his biographer, said it best when he labeled Riddell "one of the most brilliant and one of the most interesting characters in nineteenth century science in the United States."[94]

## NOTES

1. The best account of Riddell's early and middle years is his aforementioned twenty-one volume "Personal Journal" (John Leonard Riddell Papers, Manuscripts Division, Tulane University, New Orleans, La.). The retrospective entry for Mar. 4, 1834, is especially valuable, containing an informative autobiography through 1826. The only attempt at a biography of Riddell is Karlem Riess's *John Leonard Riddell*. Tulane Studies in Geology and Paleontology, no. 13 (New Orleans, 1977). It is, however, more of a source biography than a full-scale treatment of Riddell's life and career.

2. Riddell, "Personal Journal," Mar. 4, 1834.

3. Ibid.

4. For a history of Rensselaer during these years, see Ray Palmer Baker, *A Chapter in American Education: Rensselaer Polytechnic Institute, 1824–1924* (New York: C. Scribner's Sons, 1924), quotation, p. 47. The best study of Eaton is Ethel M. McAllister, *Amos Eaton, Scientist and Educator, 1776–1842* (Philadelphia: University of Pennsylvania Press, 1941).

5. "Personal Journal," May 19, 1831 (quotation), June 6, 1831. On popular science, see Hyman Kuritz, "The Popularization of Science in Nineteenth-Century America," *History of Education Quarterly* 21 (1981): 259–74.

6. Riddell's personal journals contain frequent references to his "amorous adventures"; see, for example, "Personal Journal," May 17 (quotation), June 15, 1831, Mar. 24, Apr. 3, May 20, Aug. 27, 1832, Oct. 27, 1833.

7. Ibid., May 19, 1831.

8. Ibid., May 20, 1831.

9. For a detailed list of Riddell's preparations for a botanizing trip, see ibid., June 17, 1836. On botanizing and plant sales, see ibid., July 2, 6, 10, 21, Aug. 8, 12, 19, 20, 1832. For the preparation of specimens, see John Leonard Riddell, "On a New and Effectual Method of Preserving Specimens of Organic Nature and of Obviating the Blanching Influences of Light and the Depredations of Insects," *American Journal of Science and Arts* 35 (1839): 338–42.

10. John Leonard Riddell, "The Spontaneous Vegetable Productions of Washington County, Ohio," *Marietta Republican*, July 6, Sept. 1, 7, 1832.

11. "Personal Journal," Aug. 12, 26, 30, 1832. Riddell's class lecture notes can be found in the first two volumes of his "Repository," John Leonard Riddell Papers, Manuscripts Division, Tulane University.

12. "Ohio Reformed Medical College, Worthington," in "Personal Journal," June 3, 1833. For an account of medicine, medical education, and the sectarian threat in the first half of the nineteenth century, see William G. Rothstein, *American Physicians in the 19th Century: From Sects to Science* (Baltimore: Johns Hopkins University Press, 1972); John Harley Warner, *The Therapeutic Perspective: Medical Practice, Knowledge, and Identity in America, 1820–1885* (Cambridge: Harvard University Press, 1986).

13. "Personal Journal," Jan. 27, 29 (quotation), Oct. 18, 21, 1833. As was common, Riddell was paid part of the profits, according to his entry for Oct. 8, 1833.

14. Ibid., Jan. 27, 1833.

15. Ibid., Mar. 11, May 29, June 4 (quotation), Sept. 30, Oct. 8, 9, 1833; Feb. 10, 1834. Riddell was genuinely concerned about helping his family, and even before his father's death he sent what little money he could spare home. After the loss of his father, he tried to do even more. Ibid., Mar. 7, 15, May 29, Sept. 6, Oct. 8, 1833; Feb. 12, May 11, 1834; July 14, Aug. 11, Oct. 12, 1835; Jan. 5, 1836; Mar. 13, Nov. 13, 1838; Dec. 25, 1839. After the death of Riddell's first wife, his mother came to live with him. Ibid., Jan. 11, 1840.

16. Ibid., Nov. 2, 1833.

17. Ibid., Jan. 27, Feb. 10, 28, Mar. 1, 15, 18, 19, Apr. 23, May 29, 31, July 1, 25, 27, Aug. 15, Oct. 12, 18, Dec. 8, 13, 16, 17, 30, 1833, Jan. 2, 16, Feb. 15, Mar. 2, 1834.

18. Ibid., Feb. 10, 14, 28, July 12, 26, Aug. 7, Oct. 30, Nov. 19, Dec. 16, 22, 24, 1833, Jan. 19, 20, 24, 25, 31, Feb. 1, 2, 3, 4, 5, 15, Mar. 3, 1834; John Leonard Riddell, "Observations on Showers of Meteors of November 13, 1833," *Ohio State Journal* (Columbus), Nov. 16, 1833.

19. "Personal Journal," Apr. 6, May 6, 18, June 21, 27, July 19, 31, Aug. 3, 5, 15, Sept. 9, 23, 27, 28, Oct. 25, Nov. 24, Dec. 8, 25, 27, 30, 1833; John Leonard Riddell, "Catalog of Plants Growing Spontaneously in Franklin County, Central Ohio, Excluding Grasses, Mosses, Lichens, Fungi, Etc.," *Western Medical Gazette* 2 (1834): 116–20; 154–59.

20. "Personal Journal," Jan. 8, 10, 27, 29, May 6, June 21, 26, Sept. 5, 24, 28, Oct. 2, 10, 25, 30, Nov. 19, Dec. 23, 25, 1833, Mar. 2, 5, 1834. Riddell included this species in his *A Synopsis of the Flora of the Western States* (Cincinnati: E. Deming, 1835), p. 57.

21. "Personal Journal," Apr. 1, 1834.

22. Ibid., Jan. 28, Feb. 25, Mar. 2, 4, 15, 19, 22, 1834.

23. Ibid., June 29, 1834.

24. Ibid., Mar. 15, 1834.

25. Ibid., Jan. 28 (first quotation), Jan. 31, 1834 (second quotation).

26. The standard work on Drake is Emmet Field Horine, *Daniel Drake (1785–1852): Pioneer Physician of the Midwest* (Philadelphia: University of Pennsylvania Press, 1961). On Cincinnati in the age of Drake, see Carl W. Condit, *The Railroad and the City: A Technological and Urbanistic History of Cincinnati* (Columbus: Ohio State University Press, 1977).

27. "Personal Journal," Mar. 3 (quotation), Oct. 2, Nov. 19, 1833.

28. Ibid., Nov. 9, 1833, Feb. 18, 1834; John Leonard Riddell, "Particular Directions for Collecting and Preserving Specimens on Plants, Extracted from an Unpublished Treatise on Practical Botany," *Western Journal of the Medical and Physical Sciences* 8 (1834): 18–42.

29. "Personal Journal," Mar. 29, Apr. 2 (quotation), 4, 6, 8, 1834. Riddell traveled to Cincinnati by way of the Ohio Canal. On his way there, he visited the famous Indian mounds at Circleville, writing in his personal journal on April 3:

> I walked over and examined the ancient mounds, walls, etc. The *Great Mound* is a gigantic monument of the olden time, but the people of Circleville are demolishing it as fast as they can, by cutting a street through it, or rather through the most elevated part of, for when completed as they intend, the street will pass over a considerable hill. A church has been erected near the south side of this mound, but its dimensions are truly insignificant, when compared with the tumulus on which it stands. The mound is made from the alluvian, containing pebbles, which abounds in the lands around. Now and then several feet in extent of sandy earth free from pebbles may be seen. The circular wall and ditch are sadly disfigured, and are fast falling into that ruined condition, which time and nature only could never have reduced them to.

Riddell also made use of this opportunity to observe and comment on the geology of the region as revealed in the banks of the canal and surrounding area; see his entry for Apr. 4, 1834.

30. Ibid., Apr. 10, 14, 28, May 11, 12, 1834. While Drake refused to receive Riddell in his home or to introduce him to his two daughters, he did introduce him to several potentially helpful persons. Later, relations between the two warmed again (see entries for Apr. 10, 12, 28, May 18, 22, 1834, Nov. 1, 1835, Feb. 14, 1836).

31. Ibid., Apr. 11, 14, 16, 28, May 5, 11, 12, 15, 23, June 5, 6, 16, 28, 29 (quotation), July 1, 3, 6, 13, 14, 17, 18, 19, 1834. Riddell launched this undertaking on July 5, 1834.

32. Ibid., July 1, 1834.

33. Ibid., Apr. 8, 1834; see also Oct. 18, 30, Nov. 7, 20, 28, Dec. 7, 1834.

34. Otto Juettner, *Daniel Drake and His Followers* (Cincinnati: Harvey Publishing Company, 1909), p. 186.

35. Riddell spent the period August 2–October 11, 1834, traveling through Ohio to collect specimens and to sell botanical subscriptions. "Personal Journal," Aug. 1, Oct. 18, 1834.

36. John Leonard Riddell, "Synopsis of the Flora of the Western States," *Western Journal of the Medical and Physical Sciences* 8 (1834–35): 329–74, 489–556 (this work was later published in book form). John Leonard Riddell, "Supplementary Catalogue of Ohio Plants," *Western Journal of the Medical and Physical Sciences* 9 (1836): 567–92.

37. "Personal Journal," Dec. 18, 1834 (quotation); see also ibid., Dec. 27, 1834, Jan. 2, 8, 17, 31, Apr. 5, 1835, Feb. 10, 26, 28, Apr. 3, 1836.

38. Ibid., Nov. 12, 1835 (first quotation), Jan. 27, 1836 (second quotation).

39. Ibid., Dec. 12, 1835, Mar. 19, Apr. 17, 1836.

40. Walter B. Hendrickson, "Nineteenth-Century State Geological Surveys: Early Government Support of Science," *Isis* 42 (1961): 57–71.

41. "Personal Journal," Dec. 12, 1835, Jan. 5, 22, Apr. 17, 1836; John Leonard Riddell, *Geological Survey of Ohio* (Columbus: J. B. Gardiner, 1836); *Report of John L. Riddell, M.D. . . .* (Columbus, 1837).

42. "Personal Journal," Dec. 16 1833. For Riddell's early work on the animalcular theory of disease, see his journal entries for July 25, Oct. 24, Dec. 10, 1833, Jan. 22, Feb. 5, 9, 26 1836.

43. Ibid., Feb. 5, 28, 1836; John Leonard Riddell, "Miasm and Contagion," *Western Journal of the Medical and Physical Sciences* 9 (1836): 401–12, 526–32.

44. "Personal Journal," May 29, 1836.

45. Ibid., May 20, 1832.

46. Ibid., Jan. 5, 27, Feb. 11, 14, 19, Apr. 11, May 6, 23, 29, June 2, 3, 4, 5, 6, 8, 9, 10, 13, 14, 15, 17, 18, 19, 1836. On Riddell's elopement and honeymoon-research trip, see ibid., June 22–Aug. 18, 1836. Upon her death from tuberculosis three years later, Riddell wrote of Mary:

> She was kind, tractable and pleasant in her disposition, and devotedly affectionate. She was light and elegant in her form, handsome in her features, and highly accomplished

in her education. True, as she married young and from a boarding school, she was less forward in fashionable society than many of less than half her worth; yet this retiring disposition of hers concentrated her whole feeling and attention at home, and ever but to endear her the more to me.

Despite protestations of "no disposition to let another usurp her place in my affections," two years after Mary's death, Riddell seems to have taken a mistress, Anna Hennefin, who bore him two children. On December 1, 1846, he married Angelica Eugenia Brown, twenty-one years his junior. The couple had eight children. Ibid., Dec. 22, 1839 (quotations); Riess, *John Leonard Riddell*, pp. 39, 51.

47. "Personal Journal," Aug. 14, 1836.

48. John Duffy, *The Tulane University Medical Center: One Hundred and Fifty Years of Medical Education* (Baton Rouge: Louisiana State University Press, 1984), chaps. 1–2; see also, William Frederick Norwood, *Medical Education in the United States before the Civil War* (Philadelphia: University of Pennsylvania Press, 1944), pp. 363–68.

49. "Personal Journal," Aug. 18, 1836; Riddell, "Supplementary Catalogue of Ohio Plants," pp. 567–92.

50. "Personal Journal," Nov. 2, 1836; see also, Oct. 18, Nov. 5, 8, 1836.

51. Ibid., June 6, 1838.

52. J. L. Riddell to Prof. Amos Eaton, Nov. 10, 1838, in Riess, *John Leonard Riddell*, p. 29.

53. "Personal Journal," Nov. 4, 14, 1836, May 2, 1838.

54. Ibid., July 19, 1840.

55. Riddell to Eaton, Nov. 10, 1838, in Riess, *John Leonard Riddell*, p. 29. See also, "Personal Journal," Mar. 31, Apr. 7, Aug. 26, Sept. 28, Nov. 20, 24, 30, Dec. 2, 1838, Jan. 1, Feb. 19, July 2, 1839.

56. "Personal Journal," June 6, 1838.

57. Riess, *John Leonard Riddell*, p. 60; John Leonard Riddell, *Catalogus florae ludovicianae* (New Orleans, 1852); John Leonard Riddell, "Catalogus florae ludovicianae," *New Orleans Medical and Surgical Journal* 8 (1852): 743–64; "Personal Journal," July 3, 15, 16, 17, 18, Aug. 1, 3, 5, Nov. 15, 16, 17, 24, Dec. 29, 1838, Aug. 8, 1840; For the fate of the original manuscript, see Frans A. Stafleu and Richard S. Cowan, *Taxonomic Literature: A Selective Guide to Botanical Publications and Collections with Dates, Commentaries and Types*, 7 vols. (Utrecht: International Bureau for Plant Taxonomy and Nomenclature, 1983), 4:785.

58. John Leonard Riddell, "On a New and Effectual Method of Preserving Specimens of Organic Nature and Obviating the Blanching Influences of Light and the Depredations of Insects," *American Journal of Science and Arts* 35 (1839): 338–42.

59. "Personal Journal," Feb. 22, 1841. For Riddell's tenure at the mint, see his official correspondence in Riess, *John Leonard Riddell*, pp. 82–104.

60. "Personal Journal," Aug. 24–Oct. 13, 1840.

61. Ibid., Feb. 19 (quotation), Mar. 16, 1839.

62. Ibid., Feb. 19, Mar. 6, 16, 1839, May 15, 1841, Feb. 15, 1842; "Memorial to the State Legislature for a Scientific Survey of the State of Louisiana," Mar. 12, 1855, "Minutes, New Orleans Academy of Sciences," Manuscripts Division, Tulane University, New Orleans; see also the minutes for Mar. 27, 1854. For examples of Riddell's pieces in the press, see "Geological Survey of Louisiana," *Le Courrier* (New Orleans), Mar. 8, 1838; "Geological Survey of Louisiana," *Commercial Bulletin* (New Orleans), Apr. 11, 1838.

63. For a sampling of Riddell's inventions, see "Personal Journal," May 29, 1831, Mar. 25, Apr. 4, 1832, Jan. 2, Mar. 8, 1833, Feb. 4, 7, 10, 11, Dec. 29, 1838, Jan. 24, Apr. 5, July 13, 18, 19, 1839, Dec. 23, 1840, Jan. 18, Mar. 24, June 14, July 18, 1841, Aug. 15, 1842, Feb. 27, 1843, Mar. 20, 24, Apr. 14, 1844.

64. Riess, *John Leonard Riddell*, p. 56. For Riddell's early work on the microscope, see "Personal Journal," Mar. 31, Apr. 2, 4, 8, 16, 19, 24, 1843, Feb. 8, 1844.

65. John Leonard Riddell, "Report of the Physico-Medical Society Meeting—Binocular Microscope," *New Orleans Medical and Surgical Journal* 9 (1851): 407–408; "Binocular Microscope," *New Orleans Monthly Medical Register* 2 (1852): 4; "Simplification of the Binocular Microscope," *New Orleans Monthly Medical Register* 2 (1852): 78; John Leonard Riddell, "On the Binocular Microscope," *Proceedings of the American Association for the Advancement of Science* 7 (1853): 16–22. Riddell also reported his invention to the medical press and the New Orleans Academy of Sciences. John Leonard Riddell, "Notice of a Binocular Microscope," *American Journal of Science and Arts* 15 (1853): 68; "Minutes, New Orleans Academy of Sciences," Mar. 27, Apr. 3, 1854. The best assessment of Riddell's work with the binocular microscope is J. J. Woodward, "Riddell's Binocular Microscopes: An Historical Notice," *American Monthly Microscopical Journal* 1 (1880): 221–30.

66. Woodward, "Riddell's Binocular Microscopes," p. 230. For Woodward's pioneering work in photomicrography, see Robert S. Henry, *The Armed Forces Institute of Pathology: Its First Century, 1862-1962* (Washington, D.C.: Office of the Surgeon General, Department of the Army, 1964), pp. 36–41.

67. For a cross-section of Riddell's philosophical speculations, see "Personal Journal," May 28, 31, 1831, Feb. 10, 1833, Jan. 10, Feb. 15, Dec. 18, 1834, Jan. 31, 1835, Dec. 15, 1836, Feb. 16, 1838, Feb. 3, 1839, Aug. 11, Nov. 23, 1841, Feb. 1, June 23, 1844, Apr. 6, Nov. 22, 1845, Jan. 16, 17, 19, 20, 21, 23, 24, Feb. 26, 1846, Mar. 1, Apr. 12, 1848.

68. Ibid., June 7, 1831, Mar. 1, 1833, Dec. 27, 1834, May 24, 1839, Feb. 11, 12, June 5, 1841, June 5, 7, 8, 1842, Oct. 26, 28, 29, Nov. 6, 7, 12, 1843, June 21, 1846.

69. J. L. Riddell, ed., *Orrin Lindsay's Plan of Aerial Navigation, With a Narrative of His Explorations in the Higher Regions of the Atmosphere and His Wonderful Voyage Round the Moon!* (New Orleans: Rea's Power Press Office, 1847); Riess, pp. 45–48.

70. "The Publishing Business," *De Bow's Review* 3 (1847): 587. Riddell also saw space as a place where "one might be sublimely entombed." "Prepare," he wrote, "a polished copper balloon, of sufficient size and strength, that it would when filled with hydrogen float for an indefinite period in the atmosphere say at or near the height of 6 miles, containing the corpse." "Personal Journal," June 8, 1842.

71. *Western Medical Reformer and Eclectic Journal* 1 (1847): 24.

72. "Minutes, New Orleans Academy of Sciences," Mar. 19, 1855; see also the minutes for Mar. 26, 1855.

73. "Personal Journal," Dec. 15, 1836.

74. John Leonard Riddell, "The Probable Constitution of Matter and the Laws of Motion as Deducible From and Explanatory of the Physical Phenomena of Nature," *New Orleans Medical and Surgical Journal* 2 (1846): 592–623.

75. Riess, *John Leonard Riddell*, p. 41.

76. "Personal Journal," Mar. 29, 1848.

77. Illustrative of this 'exchange is John Leonard Riddell, "Remarks on Dr. A. W. Ely's 'Examination of the Riddellian Philosophy,'" *New Orleans Medical and Surgical Journal* 3 (1846): 152–55; John Leonard Riddell, "A Reply to 'E,'" *Commercial Bulletin* (New Orleans), Jan. 14, 1846; see also, "Personal Journal," Jan. 12, 20, Mar. 1, 1846.

78. Riess, *John Leonard Riddell*, p. 6.

79. Duffy, *Tulane University Medical Center*, pp. 17, 31.

80. John Leonard Riddell, "Selected Items of Observation, Referring Chiefly to the Living Microscopic Organisms That Abound in the Waters of New Orleans and Its Vicinity, Embracing Also Some Matters Pertaining to Microscopic Anatomy," *New Orleans Journal of Medicine and Surgery* 8 (1852): 530–36, 667–69; 9 (1852): 116–19, 173–84; John Leonard Riddell, "Microscopic Observations on the Blood," *New Orleans Monthly Medical Register* 1 (1852): 98–99; John Leonard Riddell, *Introductory Lecture, on Our Knowledge of Nature, the Natural Sciences, and on Certain Truths Revealed by the Microscope . . . Delivered November 18th, 1851, Before the Medical Students, of the University of Louisiana, New-Orleans* (New Orleans: Joseph Cohn, 1852).

81. John Leonard Riddell, *Report of Dr. J. L. Riddell Upon the Epidemic of 1853* (New Orleans: Emile La Sere, 1854), p. 1. For an account of this devastating visitation of yellow fever, see John Duffy, *Sword of Pestilence: The New Orleans Yellow Fever Epidemic of 1853* (Baton Rouge: Louisiana State University Press, 1966).

82. "Minutes, New Orleans Academy of Sciences," May 1, 1854; see also, ibid., Oct. 10, 1853.

83. For the antebellum argument over the etiology of yellow fever, see Margaret Humphreys, *Yellow Fever and the South* (New Brunswick, N.J.: Rutgers University Press, 1992), chaps. 1–2.

84. Edward Hall Barton, *The Cause and Prevention of Yellow Fever at New Orleans and Other Cities in America*, 3rd ed. (New York: H. Balliere, 1857), p. 8. For a picture of the health of antebellum New Orleans, see John Duffy, *The Rudolph Matas History of Medicine in*

*Louisiana* (Baton Rouge: Louisiana State University Press, 1958–62), 2:182–83.

85. John Leonard Riddell, *Report of the Sanitary Commission of New Orleans on the Subject of City Sewerage* (New Orleans: Picayune, 1854).

86. Riess, *John Leonard Riddell*, pp. 38, 54; see also, "Personal Journal," May 22, 1849.

87. "Personal Journal," July 3, 15, 16, 17, 18, Aug. 1, 2, 1838; "Minutes, New Orleans Academy of Sciences," 1855; Riess, *John Leonard Riddell*, p. 55.

88. John Leonard Riddell, "Brief Sketch of Subjects Embraced in the Science of Botany, With Its Relation to Medicine, and Some of the Inducements for Engaging in Its Study," *New Orleans Medical and Surgical Journal* 2 (1846): 445–49.

89. "Minutes, New Orleans Academy of Sciences," 1853–1865; Karlem Riess, "The New Orleans Academy of Sciences: Its First Hundred Years (1853–1953)," *Scientific American* 77 (1953): 255–59. Riess, *John Leonard Riddell*, pp. 2 (quotation), 52–54, 66.

90. Riess, *John Leonard Riddell*, pp. 70–81.

91. Juettner, *Daniel Drake and His Followers*, pp. 202–203.

92. *New Orleans Times*, Oct. 8, 1865; see also *Daily Southern Star* (New Orleans), Oct. 8, 1865; *Daily Picayune* (New Orleans), Oct. 8, 1865.

93. Quoted in Riess, *John Leonard Riddell*, p. 34.

94. Karlem Riess, "The Versatility of John Leonard Riddell," manuscript, Riddell Papers.

# "A Long Ride in Texas"

Prairie 19 Miles West of Houston, Texas. Friday, 13th
  Sept. 1839

After much time consumed in the multifarious operation of getting ready, I at last went loose from Houston yesterday at one or two o'clock. My cavalcade cuts quite a
figure. I must enumerate and describe. First then to begin
with myself, there is my friendly self, J. L. Riddell, clad in a
buckskin coat lately taken from the body of a Cherokee
chief; with belt & Bowie knife, pocket pistols, holster pistol
and short large bore, steel barrel horseman's rifle[1]—
mounted on the tallest horse I ever saw, named Gen.
Gaines. I have saddle bags, containing clothing, money,
sewing apparatus, matches, &c. Around the neck of Gen.
Gaines and coiled, and attached to the pommel of the saddle
is a lariat, or rope of braided raw hide, fifty feet long for the
purpose of staking the General out to grass of a night. This
lariat is of essential service, and every horse on a prairie expedition must have one.[2] Next, there is Fred Banks, an intelligent boy fifteen years old,[3] bearing a fowling piece and
appurtenances[4] (to which by the way we this day owe our
dinner of wild Carolina doves), mounted on a small Spanish
steed of doubtful pedigree which has just served in the

37

Cherokee campaign.[5] Fred also bestrides saddle bags, containing clothing, ammunition &c. I have thought proper nominally to raise Fred's horse from the ranks, now that he has returned in good plight from the wars;—so I call him Corporal Trim. Lastly, a bald face gelding which has obviously seen hard service, and that too in the Cherokee campaign. I call him Captain Pluck. He is surmounted by a formidable looking pack of most heterogeneous components. A Spanish saddle tree is strongly girted on him,[6] a blanket being next his skin. A roll of blanket tied in canvas is attached to and hangs pendent from the saddle on each side. On these rest the contents of gunny bags strung across the saddle. These contents are a box of Chemical reagents, a box of tools, screws, pincers, bullet moulds &c &c, a few medicines, a box of coffee ground, and unburnt [not roasted], a frying pan, a coffee mill, crackers to eat, sugar, ham &c. Surmounting these is a portable press for preserving plants, containing half a ream of wrapping paper.[7] Upon the last is a portable tent.

The weather is fine; all is a dead level—prairies farther than eye can reach.[8] Last night I was still expecting two hours of daylight, when ten minutes elapsed it was dark.

*Euphorbia marginata,*[9] " [ditto] *maculata,*[10] *Acmella repens,*[11] *Xanthium strumarium,*[12] *Portulaca oleracea*[13] and other plants I observed on these prairies too numerous to mention in present parts.[14]

## At Mixon's, 38 Miles West of Houston. Saturday, 14th Sept. 1839

Last evening I tried the plan of crawling up within shooting distance of a deer on the open prairie. I crawled myself out of patience, say within 250 yards, and then blazed away; with the result of heartily scaring the game, for I do not think I did anything more. Herds of 3 to 10

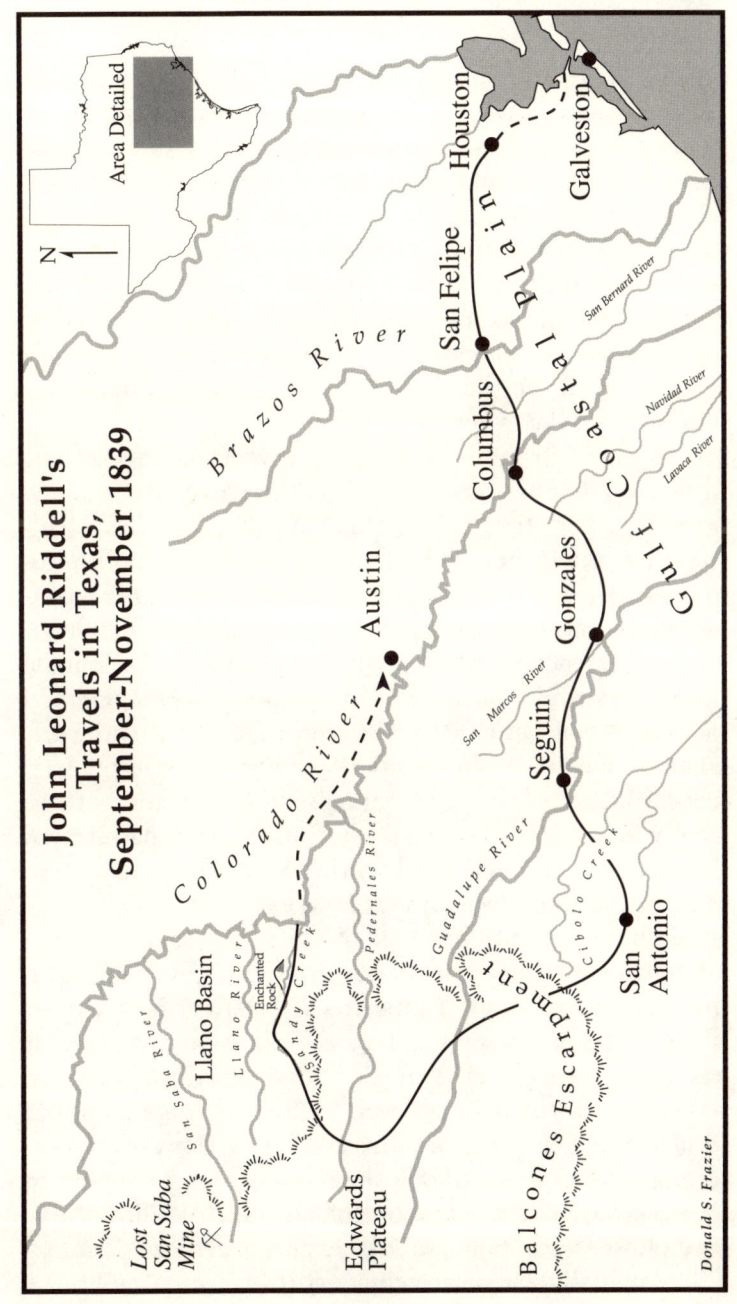

John Leonard Riddell's
Travels in Texas,
September–November 1839

Area Detailed

N

Houston
Galveston
San Felipe
Columbus

*Brazos River*

*San Bernard River*

*Navidad River*

*Lavaca River*

Gulf Coastal Plain

Gonzales

Austin

*Colorado River*

*San Marcos River*

Seguin

*Pedernales River*

*Guadalupe River*

*Cibolo Creek*

San Antonio

*Sandy Creek*

Enchanted Rock

*Llano River*

Llano Basin

*San Saba River*

Lost San Saba Mine

Edwards Plateau

Balcones Escarpment

Donald S. Frazier

39

deer we often see quietly grazing on the prairie.[15] Like other animals they prefer the vicinity of shade and water. For the last two days we have been crossing the most extensive prairie I ever saw. Level and apparently boundless as the ocean. The rotundity of the earth figure is here obvious to common perception, as evinced by the appearance and disappearance as we progress of remote clumps of trees & shrubs and other objects. The road is dusty, but otherwise fine and pleasant. Yet I see here the material of a bottomless mud and consequently impassable road.[16] Last night we camped late in the midst of the naked dry prairie. Presuming from the experience of the night previous to let the Captain and Corporal run at large with their lariats trailing while Gen. Gaines was tied to a stake near camp, we found matters rather unpromising this morning. The two subalterns not liking the dry hard grass took it into their heads to range after better, and perhaps after water. This morning the General was the only horse to be seen. When it became light enough to discern distant objects I mounted the General and with an excellent large spie-glass in hand I took a tour of observation. For once I exulted in the triumph of science, for by sweeping the horizon with the telescope I hit upon the stray animals, so far distant as to be wholly invisible to the naked eye. These horses had strayed about four miles, and not knowing which way they went, without the spie-glass I might have lost them.[17]

The universal character of the soil is sandy. It does not yet seem to be very productive. I am now writing in the shade of a clump of pine trees, constituting a timber island of 20 or 30 acres. There are occasionally dry gullies, water courses in the time of flood, now depressed 8 or 10 feet below the general level.[18] They say it is nine miles west to San Felipe on the Brassos.[19] New flowers are presenting themselves. Among them one which I suppose to be a non-descript *Solidago* or golden rod. It is diminutive corymbose, with subulate leaves, and otherwise peculiar. The shewy *Euphorbia marginata* still continues along the wayside. *Cassia chamaecrista*[20] is constantly seen. The pine is *Pinus variabilis* or *P. australis* (long-leaved pine).

Behind us is a long train of ox-wagons, bearing the archives of the Republic of Texas to the new seat of government, Austin on the Rio Colorado.[21] Mr. Mixon[22] has sunk a well about 60 feet in sand and gravel of a red color, much of it apparently good iron ore.[23]

Small mounds two or three feet high are everywhere seen in the prairie.[24]

In Texas currency, it requires a dollar a head to give the horses a feed of corn and 75 cents for a meal of victuals.[25] Today we had coffee, corn bread, fresh butter, fried beef & pork, pickled cucumbers and buttermilk. Perhaps a dozen travellers took dinner with Mr. Mixon today.

The Gama grass (*Tripsacum dactyloides*) is common in these prairies.

## On a creek or nearly dry river, between the Brassos & Colorado, near 18 miles West of San Felipe. Monday, 16th Sept. 1839.

On Saturday night we encamped on the prairie south of San Felipe. Before arriving there, and after leaving our last stopping place, Mixon's, 9 miles east of the Brassos we soon came into a beautiful rolling country, apparently productive yet principally prairie. These features in a few miles gave way to the broad level well timbered Brassos bottom lands.[26] Three miles east of the Brassos we crossed a pleasant creek, the high banks of which are thickly clad in the way of undergrowth with cane. Near this we passed two families of Mississippi planters, with 35 hands, returning to Houston from the Colorado, alleging that starvation stared them in the face at the west, as they could obtain no hog-meat, nothing indeed but corn and poor beef. It is obvious they are not Jews from their partiality to the flesh of swine. They intend returning again to the Colorado after a few months, when they will have laid in suitable supplies.

The whole country near to and west of the Brassos, probably as far as the San Saba mountains, is wickedly parched with drought. All of the smaller streams are dry.[27] I think from Austin's map[28] it must be the Rio Bernard[29] we are now encamped upon. From the width and depth of its bed I should suppose the river to be as large as the Miami in Ohio; but a feeble diminutive rill of water, discharging perhaps half a barrel of water per minute, is all that now remains of the river.

San Felipe is an indifferent collection of a dozen or so of mean wooden houses. It was much larger previous to the Texan revolution, during which it was burnt and abandoned by order of Gen Sam. Houston while the enemy were in pursuit.[30]

Passing west from the Brassos the country is gently rolling and of a prairie surface.[31] The swells are many miles in extent oftentimes so as to present a vast panorama of field and sky to the sight. Rare and remote are little clumps of trees, yet in many situations not a vestige of these can be discerned. The highest elevations are 80 or 100 feet above the lowest depressions. The grass and herbage are rather sparse, and in tracts many ten thousand acres have been scathed and blackened by fire. Hence while some portions appear green and others red, blue or yellow from prairie flowers, larger spaces in the distance perhaps appear sterile and black. The torrid sun and struggling clouds concurred also to lend a peculiar enchantment to the scene. The sky was particolored, clear and cloudy above, and the treeless earth was no less beneath for the shadows of the clouds were strongly depicted thereon as if on the object screen of a solar microscope. Then there was the effect of mirage. A dark shaded distant valley would wear the exact resemblance of a lake or water. Still more remote would appear the dark green ocean with the billows in motion. The effect being produced by undulation of heated air was more rapid than the motion of water. But otherwise the illusion was complete.[32]

We have now been 24 hours on this road and have not seen a human being. Herds of deer innumerable we have passed. Near a ravine of water we saw a great white crane.[33] The soil is sandy, and I should think rather sterile, from drought at least if nothing else. In ravines I noticed diluvial pebbles of primitive rocks, jasper, porphyry &c, none larger in size than a hen's egg. In such situation too, I notice vast beds of small nodules composed of sand indurated by oxide of iron. They are at a medium about the size of a filbert.[34] The wax myrtle (*Myrica cerifera*) is the only shrub which is constantly met with on the prairies. Burning the surface does not prevent it springing again from the roots. I have not seen it on the prairies more than two feet in height, while in the swamps of Louisiana it may be seen twenty feet high. The most common plants on the prairies west of the Brassos are *Euphorbia corollata,*[35] " [ditto] *maculata,* " [ditto] *hypericifolia,*[36] *Croton,*[37] *Cassia chamaecrista, Eryngium aquaticum,*[38] &c. Besides many new to me.

Last night our supper was corn bread and butter, and river water. This morning we made a breakfast from coffee, sugar, and fine young quails fried in butter. These we shot about the camp. On this stream are a few stinted oaks.

## Encamped on the Cibolo river, nearly equidistant, 20 miles, from Seguin and San Antonio, 22nd Sept. 1839.

We left our encampment on the San Bernard about 12 at noon, the continuing an open rolling prairie. Stopping to fix the pack we were overtaken by a dark sweaty looking footman, well armed, whom we at first took to be an Indian. Accordingly, we prepared to receive either friendly or otherwise, when upon his approach he proved to be a Dutchman who had lost his way in hunting deer. We gave him a drink of water and some imperfect directions as to his course, and he turned towards the east.

In the course of a few miles, as we neared a tract of timber in the direction of an ascending cloud of smoke we saw a couple of human beings a long distance in the road before us, one of them on horseback, the other on foot. As we neared them they both mounted the horse, one behind the other. Then leaving the road, they went a fourth of a mile from it and halted. At this time I supposed they were squaws, but by taking a peep through the spie-glass, I discovered they were two American girls, pretty good looking too; but they were dressed rather carelessly, too much as they thought, I imagine, to see company. Wishing to inquire the way, I beckoned in a friendly way manner for them to come near, but they declined. In the course of a couple of miles we came to the woods and to a rude log house where we found an old man with plenty of poultry, hogs, cattle, cheese and garden sauce [vegetables]. The old curmudgeon would in no wise part with anything, although in remuneration I offered him silver money. I offered him double the current Houston price for one of his cheeses and for some of his hens. He said he meant to take his produce &c. to Houston by & by, and then the whole would amount to something. Three miles further west he said we would come to a house and tan-yard where we might make purchases if we wished.[39] He had for some time been afflicted with the ague, and hearing Fred call me Dr. Riddell he enlarged a little upon his complaint. He had tried Dogwood bark &c, &c. I told him quinine was the thing to cure him.[40] He said he could not easily obtain it. I told him I had packed some away. He wished very much to get some if it were handy. I replied it was not near so handy to be got as one of his cheeses and that I felt myself under no particular obligation to take trouble on his account, seeing that he refused accommodating me even with the certainty of being well paid for it. He could make no reply, and so we rode and left him.

Night and moonlight soon came on, the sooner apparent as I had vainly been trying to get [a] shot at some deer. We kept our course through the heavily timbered bottom land of the Colorado, until after a seven mile ride we came at length to the banks of that river. We were ferried across on a flat boat, and took lodging for the night with a widow woman by the name I believe of Dunbar or Dunham. She kept one of the two houses of entertainment in Columbus. We took supper and the next day breakfast and dinner with her and were lodged on a good clean feather bed. She keeps the ferry. She had our big horse fed with corn during the night and furnished us with a pound or two of bread and meat for future consumption, and her bill was $8.75 Texas money.

There are rocks in the banks of the Colorado. I think they are sandstone of a light amorphous structure, although from our hurry in getting off much sooner than we anticipated I did not particularly inspect them.[41]

I intended to remain some days in Columbus until Col. Karnes[42] & party should come along, or until I should have company to travel on through this Indian country. About noon on Tuesday I met at the other public house with three Texas gentlemen who were going to start that evening, well armed on horseback, for San Antonio. We soon struck up an acquaintance and were mutually anxious to accompany each other. These were Joseph Hopkins of La Bahia, John S. McDonald, Surveyor of Bastrop & Geo. W. Reese of Bastrop.[43]

We left Columbus at two o'clock and took our way across the prairie to the south of west. This town Columbus, by the way, consists of some fifteen or twenty small houses.[44] Corn is here worth two or three Texas dollars a bushel.

Well, after some trouble with the pack, we made out to traverse about twenty five miles of rolling alternating prairie & woodland and encamped on a post-oak hill.[45] Early in the morning of Wednesday 18th Sept. we started and rode on through the same kind of country until we came to a rich wooded bottom, and the dry bed of the river Navedad.[46] We found however water for our horses in a low

45

sunken pool. I may here remark that such was the usual manner in which we met with water all along our whole route. These pools often deep are in the course of the stream, now dried up. The pond lily (*Nymphaea odorata*[47]) and sometimes the water-shield (*Hydropeltis purperea*[48]) may been seen with their leaves floating on the water. Beneath the surface grow various Algae and species of *Chara*.[49] Sometimes they contain fish, turtles and even alligators.

The water is uniformly very good. In the Navedad bottom we saw a bucket of honey hanging to a limb of a tree, but no owner in sight. We partook of it moderately and proceeded on 6 miles west of our encampment, when we came to the house of a settler whose name they pronounced Shadowen. It may be spelt Shadine. I don't know. Here we took a breakfast of corn bread, milk, coffee, a very little sugar and jerked beef fried. Price half a dollar, Texas, a piece. After leaving Shadine's we came into a beautiful region more inclined to the hilly than any we had seen. I should have sooner mentioned that all along here, to a few miles east of our last encampment, we now and then passed vacant log houses. These were deserted during the Mexican invasion, revolution, many of the owners selling out for trifles and leaving the country entirely and others subsequently settling in other parts of Texas.[50] In the course of twenty miles we passed I suppose six or eight of these vacant houses. About the point of our crossing the La Bacca,[51] which happens in a treeless prairie, the prickly pear (*Cactus ferox*, Nutt.[52]) often noticed in our course first occurs in large patches of luxuriant growth. The fruit when ripe is purple, spiney and as large as a moderate pear. The usual height of the plant here is from two to three & a half feet.[53]

At the crossing of the La Bacca, we started a drove of wild turkeys.[54] They at first flew a short distance but then preferred running, and they had to run over the naked prairie a mile or more before they got over the hill out of sight. We here saw turtle, fish, and an alligator of no mean size.

The next stream we crossed containing water was Rocky Creek which has abundance of clear running water, and from its rocky bed & banks appears correctly enough named.[55]

At length toward sundown we ascended an unusually high hill for this country, three, perhaps five hundred feet above the lowest water courses. It is a naked rather sterile prairie, completely paved with rolled pebbles, and occasionally showing the prevailing ash-colored sandstone rock. This eminence commands a grand and most extensive prospect of the country. To the right and left and behind us ridge behind ridge of rolling prairies. Before us to the west stretched out a plain of low oaken growth, blending in the distance with the horizon. While we were on the hill we experienced a strong gale accompanied with the fall of a few large drops of rain, the only rain which has occurred to us since leaving Houston. This was obviously occasioned by a dark cloudy tornado which was raging from N.W. to S.E. on our right. This ridge, as we afterwards fully learned from our host that night, Mr. McClure,[56] six miles west, has long been the highway of the Comanche Indians.[57] We observed no cross path for in fact the Indians make none, for wherever in their excursions after plunder they come to a road of the white's they separate and each one crosses singly and carefully. This Indian route was travelled two weeks since by a party of twenty or thirty mounted Comanches what boldly and in the day time [28th Aug.] took a caveard[58] of 500 horses from the immediate vicinity of La Bahia,[59] and that too under the eye of a number of white men greater than themselves. But they, the Comanches, were splendidly mounted and were splendid horsemen[60] while among the Texans there was no concern of action, and before a sufficient party could be got up for the pursuit the Indians had escaped too far with their booty.

About nine in the evening we arrived at McClure's, near the Black Creek[61] and ten miles east of Gonzales. There we fared well every way, and at very moderate expense. Mr. McClure is a very intelligent communicative man, and during the evening related a great many Indian adventures in which he had participated, or to which he was knowing. Murders had often been committed between his house and the Comanche hill. Travellers and the citizens of the country had often been chased by and made narrow escape from the well mounted savages. About a year ago he and his wife made an almost miraculous escape from eight pursuing Indians. He took his course through ravines and woods and eluded them by knowing the country better than they, for he threw them off his trail by taking water courses.

Mrs. McClure took considerable interest in my collection of plants, having acquired her taste for botany from the late Thomas Drummond who used to stay considerable at her house, as well as at Gonzales.[62] She says Drummond was a man 40 or 45, that he found a new & minute species of Cactus or prickly pear near their house.

We first saw the musquit on the eastern declivity of the Comanche hill; it was there a shrub scarce two feet high. Since crossing the Guadelupe River at Seguin, it is the prevailing tree, attaining a height of 15 or 20 feet and a diameter of trunk of about 6 inches.[63] I have remarked ever since I first saw minute pebbles near the Brassos that as I have proceeded west these diluvial pebbles become larger & larger, so that now they often occur weighing 5 or 6 pounds, being composed of porphyry, hornstone, granite, quartz &c.

At Gonzales,[64] where we arrived about 12 o'clock on Thursday the 19th Sept., we found Capt. Ross[65] with 70 or 80 men, generally mounted riflemen. As the route is considered dangerous from Gonzales to San Antonio, we considered it advisable to travel in company with Ross's men who are on their march to that place.[66] So we encamped for a short time on the banks of the Guadelupe, pronounced Wauloop, where there was tolerable grass for our horses. This is a beautiful, large, rapid river. The water is in one sense clear, is yet in mass colored a peculiar bluish white from holding in suspension, perhaps magnesium particles. It is cool and pleasant to drink. We crossed this stream by fording it at Seguin, some 35 or 40 miles above where the water has the same character. About 4. or 5. P.M. we packed up and went on the trail of Ross's company which was an hour or so in advance of us. Three or four miles above Gonzales, to the north, we crossed the Rio San Marcos, a beautiful river with remarkably bold banks, at least where we crossed it, which we effected over a high wooden bridge, altogether the most considerable one I have seen in Texas.

Gonzales, which I come near forgetting, is a small place of six or eight houses, perhaps ten or twelve. It is smaller even than Columbus. The site is fine and the lands about fertile and beautiful. This night we staid with Mr. King,[67] whose house is eight miles from Gonzales, in a direction rather northwest towards Seguin. His house is a kind of fort, very pleasantly situated.

On Friday the 20th Sept. we left King's and traversing a fine country considerably settled toward the east, we came at length to Seguin (pronounced Sageen)[68] about forty miles from Gonzales. The place contains only two or three houses. We went two miles out of our way to the Guadelupe for good grass & water. Here we encamped for the night. Next morning at 8 we again began our march, returning to Seguin and thence taking a southwest course to the Guadelupe river, where we forded it. Before coming to it, and distant a few hundred yards, I saw one of those singular immense fountains[69] which in some parts of this country give birth at once to full grown rivers. At the ford the Guadelupe with its bluish white yet transparent waters, its rocky bed, and its forest-clad banks festooned with myriads

49

of grape vines, is one of the most beautiful streams I ever beheld.[70] I suppose it is 40 or 50 yards across with a swift current and rather more than knee deep to the horses.

Immediately upon emerging from the bottom land forest, we came into the never ending musquit orchard. This musquit tree very much resembles the honey locust (*Gleditschia triacanthos*).[71] The spine is single, not trique-trous. The petiole is bifurcate, each fork being the support of a pinnate leaf; leaflets in 10 or 12 pairs; leafets linear-oblong, acutish, about an inch long; legume round, long like that of the snap bean, often striped with red, 6 inches long, seed enveloped in an adherent sweet substance or pulp. Timber the color of mahogany, heavy, durable, and beautiful. I have procured specimens of the musquit grass. I should think it a species of *Aristida*,[72] but I have not yet examined it. Its texture is almost as fine as thread; horses are very fond of it and thrive remarkably well upon it, and it is the prevailing grass where the musquit timber is found.

Yesterday about 1 or 2 P.M. we arrived at our present encampment on the Cibolo (pronounced by the Texans Seawillow). There is no running water here, the water being in standing lakes or pools, one of which near our encampment is an acre or two in extent and perfectly clear and perhaps 50 feet or more in depth. Many aquatic plants grow in it. Soft shell turtle, huge fish, some of them 4 or 5 feet long, and still more huge alligators abound in it.[73]

A mile or two back yesterday we saw a smoke two miles or so to the north. Mockasin tracks were also seen. The conclusion is there are Indians near us. When there are scattering bands of Comanches about, a large fire or great smoke is their signal of danger. At Seguin, Capt. Ross, with a detachment of 20 men, went off six miles to the north in pursuit of Indians that had just been discovered there. He has now (3 P.M.) arrived at our camp, bearing the skin of a huge bear as the only trophy taken. He found no Indians.

The soldiers and Texans universally give Indians the name John. At John's camp, look out for John.[74]

## Mission of San José, 5 miles down the San Antonio river from San Antonio. Tuesday, 24th Sept. 1839.

While encamped on the Cibolo, it is worthy of remark that my companion the excellent hunter Mr. Reese was the only individual of the whole crowd (some 80 or 90) who had any luck in killing deer. During our stay there he killed with his rifle six or seven deer. He would generally bring in the head & hind quarters and then direct the soldiers where the rest might be found. There was but one of Ross' company that pretended to have slain a deer, and as he brought in only the fore quarters, it was plain he had hit upon some of Reese's leavings. Capt. Hagler, from West Point, was often one of our mess.[75]

On the hills, about which are here a hundred feet or so above the river plains, a light ash gray limestone, intermixed with flint presents itself. It presents itself at the surface in loose, partially rounded pieces of all sizes up to a few pound's weight.

I collected many plants new to me in that vicinity. Among my old acquaintances in the way of botany, I noticed *Xanthium strumarium, Euphorbia maculata,* (?) *Virgilia lutea,*[76] *Quercus virens,*[77] *Celtis occidentalis,*[78] *Cupressus disticha,*[79] *Portulaca oleracea, Cardiospermum halicacabum.*[80] The last is universally found in river bottoms from the Brassos west.

We left after breakfast in the morning for San Antonio. The musquit orchards continue over hill and valley, with now and then an interruption of prairie, all the way to San Antonio. Two species of Yucca are often seen along. I take them to be *Yucca gloriosa* & *Y. filamentosa.*[81]

At one or two o'clock I reached the Salado river,[82] a fine running fountain born rivulet about five miles from San Antonio & twelve or fifteen from the Cibolo. The direction of this day's travel was nearly southwest. The Americans & Texans pronounce the name Salado as if spelt Salow. My company had arrived at this stream some half hour or more in advance of me, so that finding no grass there for my horses, I continued my route without delay, and with no other's company than Fred & the horses. Between the Salado & town is a dangerous pass on account of the Indians. It is most so near a high ridge about midway between them. This ridge or hill, by the way, affords a fine prospect of a range of still more elevated knobs in the direction of the San Saba mountains.[83] Well, while riding along here, I confess somewhat curious in regard to the possibility of meeting Indians. Probably more than a dozen travellers, within the last six months, have been not only robbed but murdered just thereabouts. While straining my eyes & ears to make a timely discovery of Indians, I was suddenly alarmed by a loud rattling to my right in the musquit, rather in a little valley, like persons riding furiously through the brush. I involuntarily raised my rifle, holding it with a cramp-like squeeze, when I was as suddenly relieved by the appearance of a sorrel mustang, or wild horse, running off over the hill at full tilt.[84] He was the most dangerous thing I saw fortunately.

## Mission San Josē (Pron. Hosā) Wednesday, 25th Sept. 1839.

Some twenty minutes after my alarm had subsided, a broad verdant valley opened to sight in the midst of which appeared, as if created by enchantment, the city of San Antonio de Bexar. It stands alone, surrounded by a country really held by the savages, unmarked by cultivation or improved, and at present almost sterile desert from the effects of fire & drought. It reminded me of Tadmor in the desert mentioned in the Scriptures,[85] but whether Bexar in any respect resembles the ancient Tadmor is more than I know. Having rode some 30 or 40 miles without seeing a human habitation, it was with a mixture of admiration & pleasure that I beheld before me a populous town. Descending the San Antonio river you meet occasionally with cornfields and stock farms, or ranchos, but in all other direction you see nothing in some days travel but the wild country without houses or farms.

I was especially gratified with the verdant appearance of the valley of Bexar (pronounced Bāer), chiefly because my three horses had fasted since the evening before we left the Cibolo, that is, nearly 20 hours. But upon descending through the parched brown glades and musquit thickets down to the green irrigated plains, I was sorely disappointed in finding not a blade of grass, nor a pea vine, nor anything which a horse could eat. The verdure consisted of weeds, coarse & fine, mostly bitter herbs of the natural order Compositae.[86] I crossed some of the ditches & canals for irrigation[87] and expelling a lazy drove of cattle from the shade of a spreading oak thought to encamp there. Fred & myself were proceeding to unpack, when lurking about the Alamo we saw some Mexicans curiously watching our operations. We soon saw three armed Mexican horsemen making a circuit about apparently and watching us. They soon seemed satisfied as to our intentions and sheared off. I thought likely I had trespassed by crossing one of their canals, and as there was no feed about, we packed up and passed on by the Alamo, a huge ruin of masonry,[88] and fording the clear and rapid San Antonio river entered into the City of Bexar.[89]

It is indeed to my notion a queer place. Four-fifths of the houses are thatched with a kind of reed, the cat tail flag (*Typha angustifolia*[90]) it may be, but I think it is some kind of sedge or grass. Some are in progress of erection on the Alamo side of the river, and from them I gathered some idea of the Mexican mode of building. A trench is dug around in a square for about 1 foot deep for the foundation of the walls. Timbers unhewn 4 to 6 inches in diameter, and as high as the house is designed to be, are set in this trench, side by side and on end, and bound together some 5 or 8 feet above the ground with thongs of raw hide. A few poles are attached by rawhide horizontally for supporting the roof, which is thatched with reeds. The inside is plastered with mortar, except over head and under foot. The naked ground serves universally for floor.[91] The Mexicans seem generally to cook out doors, either in the street or in their yards. The stone houses are strongly built with very thick walls and generally a nearly flat roof of earth resting on riven flat, thin pieces of timber and covered above by a coat of mortar. The Mexican inhabitants do not overburden themselves with clothing. The men work with broad hats and pantaloons on, but without a shirt. The women sometimes appear naked from the middle upwards, and the children, boys under 12 and girls under 6 or 8, go as naked as they were born.[92]

We encamped a quarter of a mile below town, in a quiet little grapevine arbor, among trees on the west bank of the river. Here we let the horses run, taking the precaution as usual to hopple them. They found no grass to eat however. I took a good bathing in the river, put on clean clothes, and marched into the town. Seeing a girl come out of a house with bread I entered therein and bought two loaves for a silver bit (12½ cts.). They were made apparently from the shorts of wheat, and both together might weigh half as much as a bit loaf in New Orleans, but in quality they were much inferior to the New Orleans bread. This bread was very acceptable to us for we had not tasted the staff of life for some days.

I at last, after some inquiry and wandering found at the northeast corner of the *plaza major*,[93] some Americans of whom I inquired about Maj. C. W. Egery.[94] I learned of Col. Patton[95] that he was at this mission. Having particular business with him, I was anxious to see him without delay. Besides I was informed the grass was good at the mission. Accordingly, I returned to camp packed and saddled up, set Fred to leading the pack horse towards the mission, rode back to town to inquire more particularly the way to the mission. Here Col. Patton introduced me to Dr. Alsbury, the principal lessee of the mission, and the gentleman in whose family Maj. Egery boarded.[96] He said I should be at home at his house—he would go down with me in the morning, and advised me to stay that night at Ross's camp a mile and a half from town where he would call for me as he passed. Having already started Fred towards the mission I was obliged to pursue him, but I was told that on my way I would most likely see Ross's camp fires. At any rate I was especially advised not to attempt to go alone to the mission at night, for it had now become nearly dark. I soon overtook Fred and we anxiously looked out for the camp fires, but saw nothing but the lambent light of a distant burning prairie set on fire by the Indians. The daring Comanches then were and now are known to be prowling in the neighborhood. Two nights previous to the time of which I am speaking they murdered a man on the same road and took his wife & two horses off as booty. And the very night of our coming down to the mission as I learned the next day, they murdered a man & stole horses two miles below the mission, making a circuit and crossing the road we travelled. As good fortune would have it, they did not come across us. Travelling at that time was an act of temerity which I did not intend, for I expected to find and join Ross's camp.

We at length arrived within the dilapidated walls of the mission, inquired with difficulty of the Mexicans the residence of Dr. Alsbury, sent for Maj. Egery at a fandango,[97] fed our horses with corn, and in the dark made a bed of blankets on the stone pavement under a portico where we slept, our horses remaining in the same walled yard. Horses here must be housed at night or they are sure to be missing in the morning.[98] A horse worth from $50 to 80 in New Orleans costs here from $5 to $12. Some say Texas is made of rawhides & Spanish horses. Bless me, they apply rawhides to more uses than we can conceive of. Rawhides constitute the carpets, chairbottoms, cots, beds, shelves, partitions, wagon beds, packs, withers, ropes, saddle-trumpery in part, and numberless other contrivances of the Mexicans. The lariat for noosing wild horses, or staking horses out to grass, is a braided rope of rawhide strands.[99]

The church here was finished in 1781. It is even now a splendid and massive specimen of Spanish architecture, profusely adorned with fine sculpture in a greenish ash-gray, fine grained, compact sandstone. The principal material of the church is calcareous tufa.[100] The floors & roofs are of plastered earth. The little wood about it is cedar, musquit, and cypress, all in good preservation. The style of architecture resembles somewhat the Cathedral in New Orleans,[101] but it is much more massive and durable. The roof is a segment of a circle, massively arched with hewn blocks of tufa. It has been robbed and may be considered a ruin, yet nothing but an earthquake or the destroying hand of man will prevent its standing unmoved a thousand years.[102]

Almas creek.[103] 7 miles West of North from Bexar.
  Sunday, 20th Oct. 1839.

Resided near a month in Bexar, where Fred and myself underwent severe fevers and where I saw much worthy of entering here. But I must defer doing it for the present.

Col. Karnes has near 120 men under his command, designed to protect the surveyors in running the line between Bexar and Bastrop counties.[104] Sent a letter home to my wife and one to Prof. Barton[105] by my namesake, W. J. Riddle or Riddell, a merchant at Bexar.[106]

Maj. Egery and Fred are with me; we have 5 horses. A Mexican whom we paid in advance to accompany us, and who was mounted on the best pony we had, ran away from us today about 10 o'clock on the road 4 miles north of Bexar under very peculiar circumstances. But those circumstances must be subject of future detail.

The country is hilly, the substratum limestone, of a pale yellowish gray color, containing flint, hornstone, and organic remains—mostly shell, some like *delphinula*. Millions of pebbles as large as one's fist strew the ground, especially on the hills. Water rather scarce.[107]

## North or left bank of the Rio Guadelupe, near 40 miles N. of W. from Bexar. Friday, 25th Oct. 1839

Our Mexican joined us on Sunday evening, having returned to town to see his woman and to get a supply of corn husks for making cigarritos, which in his drunken frolic he had forgotten at first. His name is Bustore. Many years ago the Comanches stole a child of his. He went among them to recover it and remained among them before he returned to Bexar for the span of eleven years. He then came back with his child. He has since several times been among them for different periods, from three months to three years. Within the last year he has been three months among them. He is consequently more of a Comanche than a Mexican. With them he has visited and wrought, according to his own account, various mines of silver, gold, lead &c.[108] A company of nine or ten members of equal interest, of which company I am one & Maj. Egery another member, have located lands on the San Saba river & elsewhere on Bustore's report. We expect now we will be able to visit the San Saba valley and explore the famous silver mines which we hope we have secured.[109] We have four leagues of land there, including the old San Saba fort & mission, founded near 100 years ago by the Spaniard for the purpose of working the mine. The establishment was broken up before it went into operation by a universal massacre of the whites and Mexicans by the Comanches, who claim the mine as their own property. No attempt has been made that I am aware of to revive the mission. And I believe no white man has succeeded in visiting the mine since, though several expeditions have been set on foot for that purpose. The famous expedition of Bowie and companions which resulted merely in a desperate and unequal combat with the Indians has been made known to the world by Mrs. Holley's publication.[110] Maj. Egery attempted it some eighteen months since with a party of 25 men, but just before they got to the happy valley they met a large party of Comanches and were glad to excuse themselves the best way they could and return. The chief asked the Major with apparent wonder through an interpreter what his party wanted in the Comanches' country. The Major replied that he had with him some Americans from the states who had never seen a buffalo or a bear and that to gratify them he had come out for a short time to hunt those animals. Well, says the chief, you do wrong; these are our bear and buffalo; we live upon them; they are bred on our lands. If the Comanche comes down into the white settlements and kills or drives off the cattle, then you make war upon and shoot us; so I must insist upon your immediately returning. They had to return, and the Indians escorted them, even into Bexar, but not without learning from a treacherous American of the party the real design of Maj. Egery, and it is somewhat remarkable that the Comanches have never entered Bexar since but have been exceedingly unfriendly with the whites.

The company alluded to owns 100,000 acres of land. I became a member in consideration of mineralogical services to be performed. I here own beforehand that my faith in the affair is weaker perhaps than that of perhaps any other member of the company.

On Monday night we encamped in a pleasant place after an easy day's march. The musquit tree now disappears and is replaced by live oak, post oak &c. The country becomes more & more hilly; the eminences are beautifully rounded, and the highest perhaps 400 or even 600 feet above the valleys. Land sparsely timbered, but no uninterrupted large prairies. Real thickets occur only in the *cañadas* or ravines of water courses.[111] In such places the trees and herbaceous plants are like those in Louisiana, viz., hackberry, elm, box elder, oaks, *Ilex prinoides,*[112] *Bumelia,*[113] *Rhamnus caroliniana,*[114] Dogwood, pecan, and numberless woody vines, such as grape, green brier, *Wendlandia populifolia,*[115] *Ampelopsis quinquefolia,*[116] & *cordata.*[117] I find on the high, dry places many novel, undescribed plants. My diary would be more extensive were it not that I spend most of my leisure time in drawing up descriptions of these plants from fresh specimens.

About the middle of the afternoon on Monday after we had arrived at our encampment, we saw a black bear come down from a distant hill and enter upon the plain on which we were encamped. We could not shoot the bear, for to prevent our being discovered by the Indians the Col. had ordered that no gun be fired. Some half dozen of the well mounted Mexicans of our company set off after him with swords, lances and lariats. They soon overtook Bruin and made a carcass of him. Since that time they have killed about two bears a day in the same manner. About half of them are fat and exceeding fine eating; some of them very poor.[118] That evening I was on the guard for the first time in my life.

On Tuesday we clambered over many little mountain ridges and marched through a long mountain defile, crossing the anticlinal line of the waters. As we neared Sabinas Creek,[119] a tributary of the Rio Guadelupe, the travelling became unpleasant for horses on accord of flint, hornstone, buhrstone, and rough limestone which wholly occupied the surface of the ground. The ravine of the Sabinas is exceedingly bad to cross on account of its steepness, as well as the thicket and rocks. A careless Irish servant in this ravine accidently shot a Mexican soldier through the breast. It was expected he would die, but he is now (Friday) doing pretty well and will probably recover. In consequence of that accident, we encamped early on a grassy, stony, partially oak-clad hill just on the north bank of the Sabinas. Here is an immense quantity of a clinking cinereous, cellular, siliceous limestone, or buhrstone, which cannot fail to answer an admirable purpose in the manufacture of millstones.[120] Near our camp I found a new species of *Rhus*[121] growing on the limestone cliffs. A fern grows along the margin of the Sabinas (Perhaps so named from the red cedar which abounds upon it) water, closely resembling the *Adiantum capillus-veneris*[122] of Europe; but it is probably distinct and new. Among the rocks I found the fern *Asplenium ebeneum*,[123] and near the water my *Aspidium rafinesquianum*,[124] common in Louisiana.

At one or two o'clock on Wednesday we again packed up and took our line of march for the distance of 4 miles, N by West, to the northern bank of a clear and beautiful little stream where we encamped. The land here is fine. Large cypresses grow along the water, but they looked odd to me because the Spanish moss was not to be seen pendant from their branches.[125]

Bees are wonderfully abundant in this country. The men immediately find, within 40 rods of camp, more bee-trees than they can cut down and rob.[126] So we have honey pretty plenty. But soldiers are not very liberal out of their own messes, and I find it often difficult to get a share of these little Godsends, especially as money is of no value in camp. Bread or something to eat is all that will be received.

We arrived at our present encampment yesterday (Thursday) about 3 P.M. after a march of five or six miles.

The Guadelupe is here a swift running, clear green water, say 15 yards or 20 across, and not belly deep to a horse. But there are marks of a recent rise some 15 or 20 feet higher than at present. Here about our present camp the land is rich, the soil black and deep, and the surface finely disposed with about the right proportion of woodland and prairie, and I have no doubt the region will prove eminently healthy when it shall have become settled.[127] Lying about in the distance are some high calcareous hills, possibly 800 feet above the Guadelupe.

Nothing in the way of rocks hereabouts in all these parts except horizontal limestone; the same of a cinereous hue and imbued in part, at least in many places, with silex. Organic remains, such as spiral univalve shells, favosites &c, all imperfect, are often seen.[128]

Last night we had a regular and serious rain commencing about 11 o'clock. I had prepared myself for it by stacking and covering my baggage and preparing my tent carefully. The rain fell but I heeded it not, lying dry and quiet. This morning nearly the whole in camp besides myself were soaked through and much baggage injured. It rained more or less until 1 or 2 o'clock this afternoon, and at least 25 blanket tents have sprung up on the plain, few of them finished however until the rain storm was completely over.

Our food is meat principally. I have a small store of Mexican bread preparations. One kind called *panola* consists of corn parched, mixed with cinnamon & sugar, and ground fine upon the *matat* (much like a stone slab and muller; in fact a stone slab and stone roller.)[129] Another kind called *viscocias*, made by boiling corn soft in lye of wood ashes, washing, *mattating*[130] to a pulp, as in making tortillas, and then baking dry in an oven in lumps weighing two ounces or so. Sugar is scarce. Coffee comparatively plenty. Salt plenty yet.

## Rio Guadelupe, same encampment. Saturday, Oct. 26, 1839.

We had rainy, drizzly weather last night and this morning, and it being my turn to stand guard, I fully realized the unpleasant weather.

Today a fatal accident occurred near me in camp, occasioned by carelessness in handling fire-arms. A young man by the name of Sirty, among the Galveston Volunteers, caught hold of the muzzle of his loaded rifle and attempted suddenly to withdraw it from under some blankets in his tent. The cock caught something and the rifle was discharged, the ball cutting a furrow in Sirty's right leg and right hand, entering his breast just below the heart and lodging in the spine. He fell and survived rationally an hour or so, when he died. It is fortunate that Sirty's carelessness proved fatal to no one else but himself, which it must have done if he had not intercepted the ball. He had often been reprimanded for his carelessness with arms. In fact he had no business with a cap on his gun, as it was contrary to an express order of the Colonel. His comrades buried him at sundown by a wild China tree near the middle of the plain on which we are encamped.

I have this day taken a considerable ride around our station. I observe limestone cliffs a hundred feet or more perpendicular, and a few hundred yards below, from which there are beautiful and romantic views of the swift Guadelupe and the circumjacent hills and plains.

Among the woody plants which I have seen to day are the sycamore (*Platanus occidentalis*), the redbud (*Cercis canadensis*), the slippery elm (*Ulmus fulva*), the *Ulmus alata*,[131] *Virgilia lutea, Quercus virens, Q. rubra,*[132] *Q. triloba,*[133] *Q. palustris,*[134] *Cornus florida,*[135] *Carya olivaeformis,*[136] *Rhus copallina,*[137] *Ptelea trifoliata,*[138] *Cupressus disticha, Juniperus virginiana,*[139] *Rhus aromatica,*[140] *Ampelopsis quinquefolia, A. cordata, Vitis* " [ditto][141] *Wendlandia populifolia* (is it wood? I must see tomorrow.) and many others not easily recalled to memory by lamp light. I write by the light of a rude lamp. Our Mexican hunters, every day without the use of powder, but by means of the lance, bring in a bear or two, or wild calves. In this region are large herds of wild horned cattle.[142] Buffalo I do not think we have seen yet.[143]

I spend most of my leisure in drawing up careful descriptions of plants. I do less that way than I could desire, for Fred is so unwell all the while as to be an incumbrance rather than an advantage. His lungs and some other viscera seem to be organically affected.

When I first travelled in Texas, in the Trinity country, I remarked, the wave-like appearance of the ground over considerable prairie tracts, especially on the south side of gentle hills. I had not seen enough then to solve the problem of their formation. They occur often in the Brassos & Colorado countries, and as far west I believe as Gonzales. They are the conjoint effect of drought and rain. Drought causes the surface of the ground to crack deeply at intervals of 8 or 10 feet. When it rains these cracks are filled up by surface earth contiguous to them and thus a little hollow is left, the lowest part of which points out the place of the crevice. Corresponding ridges are left of course and thus oftentimes a symmetrical and remarkable appearance is given to a surface of many thousand acres.[144] The is no musquit wood here and few prickly pears.

## 13 miles north of our last encampment on the Rio Guadelupe. Monday, 28th Oct. 1839 (3 P.M.)

Yesterday morning we buried poor Sirty, close by a tree of the wild China (*Virgilia lutea*) not oak as I mentioned before.[145] We merely marched in order with covered arms around his grave, he was let down in horse blanket, not a coffin, brush was piled upon him, then the grave filled up with earth and two rude stones placed at the head & foot. Logs were laid over the grave to prevent the wolves from digging up the corpse. The circumstance of Sirty's death made the Galveston Volunteers rather sad yesterday, and last evening they sang Psalm tunes. This morning the same funereal rites were paid to the remains of the wounded Mexican, before mentioned. He was laid beside the American, and the Mexican soldiers cut a cross on the wild China, which it is said even the Comanches will respect.

Fred seems getting rather worse. I have to help him off and on his horse, and he does not pretend to do anything. Wiederman, the surgeon,[146] thinks his lungs and bowels are seriously affected. Now he has a daily chill and a concerted fever all the while.

Our party, some of them, cut down three or four bee trees every day, and occasionally I get a real feast of honey.

The Mexicans generally lance at least one bear every day. Since our arrival at this camp two hours since, we slaughtered our last beef, a fine fat ox. Henceforth we *must* live by hunting, or on our horses and mules. In a day or two we hope to come in with droves of buffalo.

Last evening about 9 considerable alarm was produced in camp among the knowing ones in consequence of an unusual howl, some like that of the black wolf, which was thought to have been produced as a signal by a Lywookany or Comanche Indian. Col. Karnes in consequence ordered the horses removed from their station above us and brought as usual into camp. Some 25 men guard them constantly. Finally it became the belief that the howl was produced by a large white wolf indigenous to the San Saba mountains, called loover or looper or luper by the Mexicans (From *lupus Lat.* no doubt.) It is said to be twice the size of the large black wolf, or three times as big as the prairie wolf, and sufficiently bold and strong to destroy a horse.[147]

We are in a rich and beautiful valley, well watered on both margins by living streams large enough perhaps for grist-mills. Limestone cliffs of horizontal strata often show themselves, some 40 or 60 feet high, in the adjoining hills. Prairie and timber well and properly intermingled. Soil of this rather roiling plain, deep, rich, dark, and easily tilled. Timber live oak, post oak, wild China, musquit &c. Who could ask for a better site for a stock farm. Most of this plain I have no doubt can be artificially irrigated.[148]

Mr. Lindsey, the Surveyor of Bexar county, who is now with us running the line between Bexar and Bastrop counties, calls the creek east of us Line Creek.

Today we met with plenty of black haws, ripe and sweet, the only edible fruit which has as yet occurred to us since we left Bexar. This fruit is produced by a shrub or small tree known to botanists under the appellation *Viburnum prunifolium.* I see the button ball in these parts (*Cephalanthus occidentalis.*)[149] We occasionally see flocks of ducks and wild geese and the strange sand-hill crane. Its notes are the most wild unearthly sound I ever heard. It is I believe much larger than the blue crane common in New York.[150]

Encamped on the right or south west bank of the Rio
    Perdinalis,[151] near 65 miles north about 19° west
    from Bexar. Thursday, 31st Oct. 1839.

On Tuesday morning at an early hour for us our Expedition got in motion. During this day's ride of 17 miles, we crossed a range of limestone mountains, dividing the waters of the Perdinalis from those of the Guadelupe. This range consisted of ridges, knobs, ravines & defiles, the horizontal, cinereous, strata of limestone being the material of construction. The prospect from some of the knobs we had to cross was grand and extensive. I suppose they might have been 1000 feet above the Guadelupe or Perdinalis.[152] They are partially clad in a rather stunted growth of black jack and other oaks. I found here a new species of *Senecio*,[153] which from the pleasant odor of the flowers, I have denominated *Senecio fragans*.

Tuesday night we took possession of a beautiful old Indian camping ground on the Perdinalis. The Perdinalis bottom (!) lands are pretty level, dry, exceeding rich and abound in musquit and oak & wild China timber, intermingled beautifully with grassy prairies. The Perdinalis is now a smart, clear, pebble bottom creek, but there are marks of a recent rise fifteen or twenty feet above its present level. It abounds in catfish, perch, sunfish &c. Such at least are the names which our folks give them, though it is quite likely some of them are undescribed species. We still have a bear or two every day, but bee-trees seem scarce upon the stream.

Yesterday some 20 or so went out to hunt buffalo, of which we saw plenty of fresh signs. The intention was to lay in a stock of jerked meat for future consumption in journeying to and within the San Saba valley. But the hunters returned about 3 P.M. without luck, so we were ordered to pack up and march. We came west some five miles up the Perdinalis.

Fred is so weak and unwell that he rode with some difficulty. I gave him the easiest going horse.

Shortly after we encamped last evening, the hunters went out and slew a fat large bear, and a bull buffalo in tolerable order. The beef of this buffalo is pretty good. Last night a shower came on to the great annoyance of those who had no tents, that is to all except myself. Today a great many blanket tents have been built for it rains more than half the time.

Wild turkeys are very abundant, and far from being as wild as in Eastern Texas. It is said the Comanches do not hunt or eat them.

Musquit, post oak and black jack are the prevailing trees in this vicinity.

Encamped on a rocky clear creek among flinty, limestone knobs, the creek running N.E. some say into the San Saba,      miles from      . Tuesday, 5th Nov. 1839.

### Friday last & Saturday

We left our encampment at an early hour and marched through a beautiful, arable, well watered region that day; first 4 miles N.E. to rejoin the Comanche trail[154] we had left two days before; next 10 miles N.W. up the valley of a small creek, among high picturesque knobs, towards the anticlinal line. Here we were overtaken by Thos. Dennis, one of the party of surveyors, running his horse at the utmost speed. He announced that Hays and the spy[155] Antonio Perez (who had left the party of surveyors to visit a supposed silver mine some 20 miles off) had discovered a large body of Comanches marching on horseback towards the southeast. They supposed the Comanches had not in turn discovered them. Soon the whole party of surveyors came up at their best speed. The Indians were seen some 15 miles S. E. from the spot where we then were. After a few minutes delay the expedition was set in motion on its back trail, and at the distance of ten miles where we took the turn we were ordered to encamp. Immediately, that is, within 20 minutes thereafter 63 of our men, well mounted and armed and divested of everything superfluous, went in pursuit of the Indians. As Fred was very sick I concluded to remain and help guard the camp. Among the 63 were some 12 or 15 Mexicans besides Old Dumacia the head spy. They found and followed the Indian trail without difficulty until it became dark. It was very dark, as there was a sky full of clouds and no moon. Old Dumacia took the lead, every now and then jumping down from his horse and lighting a lucifer match to see if he were right.[156] Thus they followed on until 11 o'clock. Then they encamped until day began to dawn, when they again set off in pursuit. They had proceeded but a few hundred yards when they came at once upon the Indian encampment. The Americans were ordered to dismount, tie their horses and make the attack on foot; the Mexicans were to co-operate as mounted cavalry. The attack was boldly made by the Americans; 6 Indians were killed by rifles and one prisoner taken. Some 30 or 40 horses and mules, all of the saddles, bridles, buffalo robes and blankets and most of the lice and arms of the Indians were among the booty. There were twenty-four Indian warriors in this gang, and it appears they were on a marauding expedition destined for the white settlements. The conduct of the Mexicans in this engagement deserves an especial notice. Two or three of the Indians escaped early in the battle on good horses which they had the night before tied up. The rest, near 14 in number, had to run away on foot and nearly naked in fact. If the Mexicans had done their duty every one of the 14 might have been cut off. But instead of pursuing and fighting the Indians, they immediately set about securing the most valuable part of the plunder, such as wallets, robes, arms &c. This was of great disadvantage in another way, as in consequence several Indians were permitted to escape by the Americans under the impression they might be their present allies the Mexicans. The fact is the Mexicans have an aversion to killing the Comanches, for by so doing there is danger of murdering their relatives. The prisoner taken was the son of a Mexican woman by a Negro father.[157] He was born in Bexar and really had a cousin Herman and some other relatives among our Mexicans. His mother when he was very young was taken by Comanches, among whom she afterwards lived long as an Indian wife and died. The prisoner was a perfect Comanche in habits, though not in spirit it seems for it is said no true Comanche will ever suffer himself to be taken prisoner. Our prisoner was mortally wounded by a shot through the breast and back, yet he conversed freely in Spanish and seemed generally at his ease. Sometimes for hours pain would make him groan audibly. This morning about daybreak, seeing that his end was approaching, he arose on his feet and chanted aloud his Indian death song for a minute or more, lay down, and in ten or twelve minutes silently expired.

A division of the Indian plunder was attempted, but the thievish Mexicans from Old Gonzales, their Commander, down concealed away the most valuable part of it. The Americans were so exasperated by their palpable trickery that they with one voice declare that in the next engagement they will not be very particular as to who is a Mexican or who an Indian. For instance, out of 24 lariats taken the Mexicans concealed all but two. They retained 7 or 8 bows with quivers and a dozen or so of dressed hides. And what is the worst of it, Col. Karnes, expecting place and power from the Federal Mexicans, has protected them in their pilferings.

Sunday.

This morning we marched N.E. a few miles in order to take the back trail of the Indian party. Some fine land— prairie and low oaks as usual.

Monday.

Left our encampment and marched perhaps 10 or 12 miles north across a high, flinty ridge which proved to be the dividing ridge for the waters. These mountains are covered with a low growth of different oaks, live-oak, pin oak, black jack, &c. No prairies, but the trees are very scattering and never push up more than 20 feet high. Soil full of flints and hard limestone fragments. Limestone rests horizontal with occasional small beds of red marl, and sandstone, greenish blue, and other red, resembling Eaton's saliferous rock formation.[158] Radiated quartz occasionally seen.

Tuesday.

Where we are water fine, plenty of game, but little land free enough from stones for easy cultivation.

(2 P.M.)

Since writing the above we have removed our encampment about two miles to the north, on the north bank of a fine running picturesque creek.[159] In places impending walls of limestone and red & blue marl show themselves in the hills fronting on the creek. This country is more sparsely wooded than the apple orchards in the northern states. Large game can consequently be conveniently seen at a distance and if not too fleet, run down on horseback. The soil looks light in texture, dark in color and rich, and I have no doubt, though often rather stony, is well fitted for cultivation. From this remark I must except the most elevated and flinty knobs. This region I am of opinion must be elevated perhaps 1500 or even 3000 feet above the level of the sea, possibly more. It is the general opinion that rain often falls here during the summer when there is a continued drought about Bexar and the low country.

Yesterday permission being given to some of the Americans to hunt with their rifles, they soon brought in 11 deer. The Mexicans killed also a buffalo near our present encampment.

Why we travel so little, start so late and encamp so early, I am a loss to devise. We shall certainly never break down our horses at this rate.

I discover hereabouts some primitive bowlders of small size and large angular fragments of a hard reddish brown quartzose rock, resembling in structure Eaton's granular quartz.[160] I think on the whole the limestone strata must dip slightly to the south and that we are now coming upon older formations.


## Still on the Banks of La Petit Creek, or Lapeta Creek (the which, before I learnt the old name, I had thought to call Kaolin Creek,) two miles below our last encampment of Tuesday. Saturday, 9th Nov. 1839. (12 M.)

Our commander Col. Karnes has been afflicted since Tuesday last with an attack of fever. Hence we make no progress in our expedition. So at least it is wished we should believe. Karnes is certainly unwell, but through some of the surveyors I happen to have a peep behind the curtain. Many years ago Old Dumacia, the head spy, showed to James Bowie a mine of silver, it is said, not more than 5 miles from our present encampment. Karnes, probably as agent for a company with which he is connected, made application a year since for a league of land about the mine, calling in his application for certain supposed rocks &c. After we had encamped on Tuesday, Hamm

Hays, the Deputy Surveyor in favor with Karnes, and Old Dumacia set off to find the mine. Dennis, Van Ness, Egery & Bustore and I believe some others set off with them unasked. Old Dumacia led them a random route over the mountains, travelling some 25 miles to make 5 miles of headway. Said he could not find the mine. It is now understood that he would not then go to it. For it is death and damnation by the Comanche law to show a mine of precious metals to more than one. Today Hays & Dumacia have set off again in quest of the mine.

There is a decided and important change of surface rocks in these ravines.

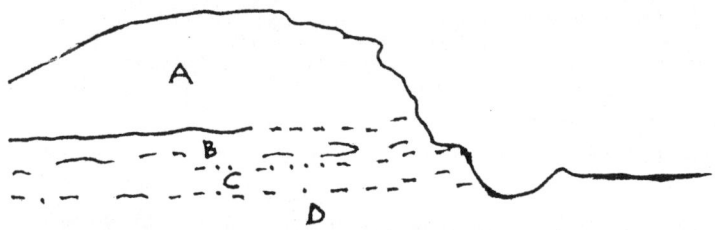

The best exposure of rocks I have seen is near last encampment, 2 miles up this creek.

There A = cinereous limestone, as heretofore.

    B = limestone, interspersed with beds, often extensive of red and bluish green marl.

    C = primitive granular quartz, color mostly brownish red.

    D = granite, mica golden colored mostly & black, feldspar flesh-colored.

In the bed of the creek, where the granular quartz may be seen overlying the granite, the strata of quartzose rock dip to the Southeast about one foot in 25 or 30. There in the bed of the same creek I think the dip of the quartz rock, the only one seen, is rather greater in the same direction. The granite appears destitute of seams or strata, but is traversed in various directions by veins of transparent quartz.

On the limestone hills about I picked various forms of calcareous spar, radiated and druzy quartz, yellow & red carnelian, variegated agate, and every variety and color of flint and hornstone.

In the granite various minerals are discernible, a part of which I will here enumerate after having visited some localities which lie in our future route. In the granite bed of the creek above mentioned the component parts of the rock cohere loosely, and often the feldspar has disintegrated into a fine white clay—the same used by the Chinese in the manufacture of porcelain, under the name of kaolin; whence I was inclined to name the stream Kaolin Creek.

I believe the Americans and Mexicans of our party hate each other most cordially. This was developed on the part of the Americans immediately after the Mexicans had barefacedly stolen and concealed the best of the Comanche plunder. Night before last one of old Col. Gonzales' box was broken open and robbed of its contents, bread and Mexican sugar. (In little loaves wrapped in leaves, and called *potonzes*.) I believe, and so does everybody else, that it was done by some American of Capt. Wilson's company. We are all heartily ashamed of the act. But no measures were set on foot to detect the guilty.

The land is rich and the water pure and plenty near us.


On the banks of La Peta Creek (It is that creek, which
    Mr. Lindsay now doubts) 10 miles or so down East
    from Karnes' encampment of Saturday last. Tuesday,
    12th of Nov. 1839.

I am now in company with the Surveyor's party only. We left Karnes' company on Sunday morning with the intention of completing the county line to the Colorado and of thence proceeding to Austin.[161] We left with a party 18 or 20 strong, namely, William Lindsey, Surveyor of Bexar County, Bexar:[162]—James B. Collinsworth, Depty.

Surveyor, Bexar, John C. Hays, Depty. Surveyor, Bexar:[163]—
John James, Depty. Surveyor Bexar:[164]—John P. Cockrill,
Matagorda, Thos. M. Dennis, Matagorda; Joshua Thread-
gill, Bexar,[165] F. L. Pascal, Bexar,[166]—L. N. Nobles, Bexar,
George Van Ness, Bexar,[167] Moses Hessken, Bexar:[168]—
Thos O. Moody, Matagorda, William E. Roberts, Galve-
ston, Fred Banks, New Orleans, Maj. C. W. Egery, Bexar,
Bustore, an old Comancheized Mexican, Amsilmo Fer-
nandes, Whan [Juan] Fernandes, Mexicans, and Frank, a
black man—making twenty individuals with myself.

On Sunday about 8 in the morning we took leave
of our acquaintances in Wilson's and Karnes' party with
a view of encamping that night near an immense dark
granitic, rounded rock. Immediately after getting across
the first range of limestone mts. we saw two miles distant
to the left in the vast valley before us a smoke arising from
a burning prairie. Indians were at once suspected, and it
was thought prudent to send a messenger to Karnes for 50
men or so to explore the matter thoroughly. They came in
an hour or so and we rode to the place and found nothing
but a few mockasin tracks.

(5 P.M.) Encamped on the same creek, some 5 miles
  lower down.

It is now currently believed in our crowd (I use a
Texan expression) that this fire was set to the prairie by
the secret order of old Col. Gonzales, Captain of Bexar
Volunteers in order to terrify our party with the fear of
Indians, that they might return and accompany him to the
Rio Grande. As near as I understand it, it is the intention
of himself and Karnes to take the town of Presidio de Rio
Grande,[169] now under the control of the centralists, and
therein they expect to find booty and thereabouts many
herds of cattle & horses. Old Gonzales seems to be afraid
to return to those parts without considerable force.[170]

After again separating from the soldiers we proceeded to our preconcerted place of encampment. I must not here omit to notice the most grand and extensive prospect, which is commanded by any of the highest mountains about the valley in which we encamped. This I first remarked in crossing the limestone mountains after leaving Karnes' camp. Looking to the northeast, hill behind hill, range behind range, all lying beyond a valley of several leagues in extent, fade at length into a faint blue not distinguishable from the sky. On the left lies a naked, dark and particolored granitic range of knobs, the highest equal any in elevation the limestone mountains to the right. One of these granitic knobs is said to be called by the Indians the Enchanted rock.[171] It is perhaps 400 feet high, and is a pretty regular hemisphere in form like an apple half protruding from and half-buried in sand. Its dark color is owing to lichens which grow upon it. From our hurried movements I was unable to attempt its ascent.

## Near 7 miles still lower down on the same creek, in a direction East. Wednesday, 13th Nov. 1839. (5 P.M.)

We are still in a country where the various crystalline primary rocks prevail. We have blocks and continuous masses of granite hornblende rocks, gneiss and various primary slates, as mica slate, talcose slate.[172] Yesterday our course lay through the narrow rock bound ravine of the creek; some miles of the distance we found it only practicable to travel in the granitic sand of the creek bed. On both sides there were high hills capped with the cinereous limestone dipping slightly to the Southeast; near the creek it graduated into a kind of quartzose altered secondary rock intermediate in appearance between primary and secondary rocks. I have today seen some of the largest masses of rock crystal (hyaline quartz) which I ever saw, the same lying partially concealed by the soil. There appears to be considerable quarries of it.[173] I also saw similar quarries of feldspar of a pale yellowish white. I should suppose it would be well adapted for the manufacture of porcelain. Unquestionably with time to search and means to transport, an extensively diversified cabinet of minerals might be taken from these regions.

Musquit timber still abounds though rather less plenty than around San Antonio. The land in the valleys and on the less rugged hills is of a fine quality, light and loose in texture and fitted for easy cultivation.

Day before yesterday, Roberts, a soldier discharged on account of bad health, and Frank the colored man, whom I have engaged to serve me, were left behind, and they did not rejoin us until yesterday morning. Frank was lost thus: His lariat was used about 2 P.M. to pull down a bee tree, wherein we found more good honey than we could all eat. Somehow his lariat was left behind, and he contrary to my advice returned near half a mile for it. At this time I was behind and out of sight of the others, having been detained with Fred. As the route lay up, down, and repeatedly across branches and the main creek and over dry stony hills, I found it very slow work to follow the trail myself. At length I overtook the party, and they soon made a halt to regulate the pack of one of the mules. Roberts had so engorged himself with honey as to be very sick. He lay down. When we started we roused him and he went to his horse, and it seems before he mounted we were out of sight, having changed our direction and crossed the creek. He could not follow our train and when Frank came along on the trail he heard a gun fire to his right, and half a mile from the trail he found Roberts perfectly bewildered. He got him slowly along, and camping when it became dark they reached our place of encampment about 9 yesterday morning.

This stream is now small barely enough to carry a good grist mill, but it is plain to see that it sometimes rises 30 feet, and has then an average width of ¼th of a mile. The sandy bed is now about 100 yards wide.

I see fewer fountains than I should expect.

On the East or left bank of a large river, which most of us
take to be the Colorado, some few the Rio Llano
(Pron. Yano), perhaps 8 miles rather north of east
from our last night encampment. Thursday, 14 Nov.
1839. (4 P.M.)

After encamping yesterday we busied ourselves with
cutting up and drying on a scaffold of sticks the fine fat
beef of a buffalo cow, which Old Bustore had run after and
slain with arrows and pistol balls near the spot. I never ate
finer beef in my life. Yesterday for the first time we actually
came in with herds of buffalo. I suppose the first here we
saw contained 500 individuals. It was from this drove that
Old Bustore on a swift Comanche horse,[174] selected our
first cow. I assisted in running down and shooting a buffalo
calf, but it had been long wounded by an Indian arrow and
was consequently not eaten by us.

We have today passed over a fine country with here
and there high obtusely pointed hills. The first part of our
ride was over primitive slates and the soil formed by their
disintegration. The mica slate of a dark gray color is most
abundant. The strata dip at an angle of 30 or 40 degrees in
various directions, though constant in direction over considerable areas. Hornblende slate I saw, and a close approach, in the bed of the La Peta creek where we crossed it,
to argillite or roofing slate.

Near some falls of the Llano or Colorado river (whichever
it proves to be). Friday, 15th Nov. 1839

(7 in the evening—moonlight)

I have fitted up a gigantic candle from wrapping paper
and buffalo tallow. It answers well.

The surveyors this morning established the county
line upon this stream.[175] It is 6 or 8 miles above our present
encampment. Before leaving this morning we found and

robbed a bee tree. I suppose a skillful bee hunter might find at least 10 bee trees every fair day. The bees generally in this country live in the hollow post oaks and the orifices to the honey are seldom above the reach of a man standing on the ground. We take nearly one swarm a day, and have nothing but my old dull hatchet to enlarge the orifice so as to get out the honey, and lariats, perhaps to pull and rive the tree asunder. On an average each tree yields 3 gallons of honey at this season. It proves always a delightful treat to us all, especially as we have for some time been out of sugar. All the articles of food consumed for which I am indebted to commerce or to civilization are coffee and salt. Some of our party still have a little flour; yet all mainly live on game with salt and coffee, and all with the exception of Roberts and Fred, who left Bexar unwell, enjoy excellent health. This kind of life has great charms.

Today we killed several turkeys, rather poor, a very good buck, and a bull buffalo. Game is exceedingly plenty. We saw in our route today three or four dead buffalo, said by Old Bustore to have been killed by the great white wolf.[176] Passed over many naked granite eminences. One to the left was some square miles in extent. Later we passed over higher secondary limestone knobs. Soil excellent. Musquit and oak timber, with cotton wood (*Populus laevigata*[177]) and ash on the river bottoms, intermingled with Pecan and Sycamore.

Here at the falls the strata dip near 10° a little to the south and west.

A = black variegated—10 f.
B = marble—black fissile slate 3 f.
C = fine hard variegated black marble. 10
D = grey crystalline marble

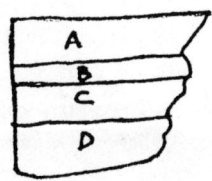

This is a better diagram.
And the place observed is in
the bed of the river.

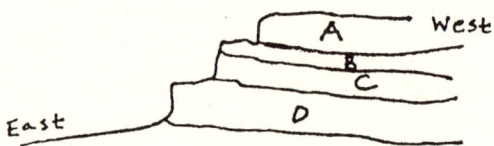

Some of our party have followed the river a mile or so down and say that the total fall cannot be less than 100 feet.

The strata C & D will rank as fine varieties of marble, for the blocks can be obtained large enough entire for the pillars to churches. The structure though diversified with greens and darker cloud, is yet wholly continuous and strong, throughout the block.

D approaches in structure by being crystalline to primitive marbles. I observed so.

NOTES

1. A short-barreled, large-caliber rifle designed for use on horseback. For a contemporary account of such a weapon, see Louis A. Garavaglia and Charles G. Worman, *Firearms of the American West, 1803–1865* (Albuquerque: University of New Mexico Press, 1984), p. 119. See also, *Texas in 1837: An Anonymous, Contemporary Narrative*, edited with an introduction by Andrew Forest Muir (Austin: University of Texas Press, 1958), p. 55; Ferdinand Roemer, *Texas: With Particular Reference to German Immigration and the Physical Appearance of the Country*, translated from the German by Oswald Mueller (San Antonio: Standard Printing Company, 1935), p. 117.

2. As Riddell notes, the lariat, a corruption of the Spanish word *la reata* (rope), was invaluable on the Texas frontier. For its history and uses, see William MacLeod Raine and Will C. Barnes, *Cattle* (Garden City, N.Y.: Doubleday, Doran & Company, 1930), Appendix I; Philip Ashton Rollins, *The Cowboy: An Unconventional History of Civilization on the Old-Time Cattle Range* (New York: C. Scribner's Sons, 1936), pp. 137–42.

3. Frederic Banks was the cousin of Riddell's deceased wife and lived with him. John Leonard Riddell, "Personal Journal," Dec. 22, 1839, John Leonard Riddell Papers, Manuscripts Division, Tulane University, New Orleans, La.

4. A shotgun and accessories, largely a shot pouch, powder flask, caps, and wads. Garavaglia and Worman, *Firearms*, p. 69.

5. Riddell is referring to the so-called Cherokee War of 1839, in which the East Texas Indians were forcibly expelled from the Republic of Texas. This campaign was also apparently the source of Riddell's buckskin coat. "The Expulsion of the Cherokees from Texas in 1839," in John Henry Brown, *The Indian Wars and Pioneers of Texas* (Austin: L. E. Daniel, [1890?]), pp. 66–69.

6. A saddle tree is the frame of the saddle. For its function in early Texas, see Frederick Law Olmsted, *A Journey Through Texas; Or, A Saddle-Trip On the Southwestern Frontier* (New York: Dix, Edwards, 1857), p. 54; *Texas in 1837*, p. 55.

7. The press was to be used to flatten plants; the wrapping paper, which Riddell elsewhere described as "soft, bibulous paper," was to preserve them in. These were essential items for the natural historian. Riddell, "Particular Directions for Collecting and Preserving Specimens of Plants, Extracted from an Unpublished Treatise on Practical Botany," pp. 18–42.

8. Riddell was deep into the Gulf Coastal Plain. Occupying the southeastern and southern portions of the state and bounded on the west by the Balcones fault zone, which ran roughly along a line from Del Rio through San Antonio, Austin, and Temple, it was one of Texas' largest physiographic provinces. In its native state, this coastal prairie region was grassland, broken only by hardwoods along the water courses. Terry Jordan, with John L. Bean, Jr., and William M. Holmes, *Texas: A Geography* (Boulder: Westview Press, 1984), pp. 7, 14–16, 31. Riddell's perception of this physiographic feature's monotonous flatness was a commonly shared perception of early settlers and visitors. Marilyn McAdams Sibley, *Travelers in Texas, 1761–1860* (Austin: University of Texas Press, 1967), pp. 46–51; Roemer, *Texas*, p. 57.

9. Snow-on-the-Mountain. Plant identifications are derived from Thomas Nuttall, *The Genera of North American Plants*, 2 vols. (facsimile of the 1818 edition), Introduction by Joseph Ewan, (New York: Hafner Publishing Company, 1971); André Michaux, *Flora Boreali-Americana*, 2 vols. (facsimile of the 1803 edition), Introduction by Joseph Ewan (New York: Hafner Press, 1974); John Torrey and Asa Gray, *A Flora of North America* (facsimile of the 1838–43 edition), Introduction by Joseph Ewan, (New York: Hafner Publishing Company, 1969); *Gray's Manual of Botany: A Handbook of the Flowering Plants and Ferns of the Central and Northeastern United States and Adjacent Canada*, 8th ed., largely rewritten and expanded by Merritt Lyndon Fernald (New York: American Book Company, 1950); John T. Kartesz and Rosemarie Kartesz, *A Synonymized Checklist of the Vascular Flora of the United States, Canada, and Greenland*, (Chapel Hill: University of North Carolina Press, 1980); Harlan P. Kelsey and William A. Dayton, *Standardized Plant Names*, 2d ed. (Harrisburg, Pa.: J. Horace McFarland Company for American Joint Committee on Horticultural Nomenclature, 1942); Donovan Steward Correll and Marshall Conring Johnston, *Manual of the Vascular Plants of Texas*

(Renner, Tex.: Texas Research Foundation, 1970); Stephan L. Hatch, Kancheepuram N. Gandhi, and Larry E. Brown, *Checklist of the Vascular Plants of Texas*, Miscellaneous Publication no. 1655 (College Station: Texas Agricultural experiment Station, Texas A&M University System, 1990). Riddell made frequent mistakes in the spelling of plant names. These have been silently corrected. Where the nomenclature has changed, the modern form is given in parentheses. Finally, as illustrated above, the common name of each plant is given.

10. (*Chamaesyce maculata*); Spotted Spurge

11. (*Spilanthes repens*; a variety of *Spilanthes americana*); Creeping Spot-Flower

12. Cocklebur

13. Purslane

14. The profusion of wild flowers on the Texas prairies was a source of amazement and fascination for many travelers in the region. Sibley, *Travelers in Texas*, pp. 46–47.

15. The deer was the most common game animal in early Texas. Travelers were consistently awed by their numbers. Ibid., pp. 57– 60; Roemer, *Texas*, p. 83.

16. During inclement weather Texas roads were the bane of residents and travelers alike. Roemer, *Texas*, pp. 66, 69–70.

17. The problem of horses straying was common. Roemer called it one of the "'petites miseres' of traveling in Texas." Ibid., p. 176.

18. Riddell is describing the alfisol that makes up much of the Gulf Coastal Plain. Early settlers eschewed these thin-soiled prairies for the highly regarded alluvial soils of river-bottom prairies. Jordan, *Texas: A Geography*, pp. 37–39. To Roemer, this general region was "a rather characterless hilly country, in which open prairies and oak forests alternate." "Sandy, infertile pine-covered hills," he reported of the area of the Colorado River, gave way to "the rich level bottoms of the river." Roemer, *Texas*, p. 165 (first quotation), p. 166 (second quotation).

19. Brazos River

20. *Senna* genus

21. Riddell is reporting the transfer of the Texas capital from Houston to Austin. The capital experienced a nomadic existence between independence in 1836 and 1853, when it was permanently located in Austin. Seymour V. Connor et al., *Capitols of Texas* (Waco: Texian Press, 1970).

22. Although six of the Republic's thirty-two counties are missing and the returns from others are incomplete, the best single source for specific names in early Texas is the compilation of the 1840 tax rolls. See Gifford White, ed., *The 1840 Census of the Republic of Texas* (Austin: Pemberton Press, 1966). Mixon does not appear in this listing.

23. While the Gulf Coastal Plain does not contain one of the major concentrations of iron in Texas, the region, beginning with the Cretaceous Period and lasting until the end of the Eocene Epoch of the Tertiary Period, experienced a long period of deposition of iron minerals. E. H. Sellards and C. L. Baker, *The Geology of Texas*, volume 2:

*Structural and Economic Geology,* University of Texas Bulletin no. 3401 (Austin, 1934), pp. 422–82, especially, pp. 453–58.

24. These mounds were apparently made by the Texas leaf-cutting ant (*Atta texana*), distinctive for the large mounds of dirt marking the entrances to their nests. William Steel Creighton, *The Ants of North America,* Bulletin of the Museum of Comparative Zoology, no. 104 (Cambridge, 1950), pp. 329–31. Further light was shed on these unique ants several years after Riddell's observation of their mounds by one of his students in a communication Riddell had read to the New Orleans Academy of Sciences:

> Prof. Riddell presented a communication from Mr. E. M. Walker, of Yorktown, DeWitt County, Texas, . . . who is now in the Medical Class of the University of La. on the subject of the mound-building ants; together with a box containing specimens of the Ants themselves. The communication was read to the Academy by the Rec. Secretary; and, on motion of Messrs. Riddell and Lindsay it was ordered to be spread upon the minutes.
>
> The following is a copy of the paper which was in the form of a Diary. . . .

> February 11th. 1853. In travelling near Yorktown, Texas, in Lat. N. 27 ½°, I discovered various hillocks, which, upon examination I found to be what the old Texians call "Ant-Villages";—inhabited by a peculiar ant, a specimen of which accompanies this Paper. In March 1853 I settled in Yorktown, and had many opportunities of witnessing the habits of these ants: which opportunities I improved. From various observations in the vicinity of the above-named place, as well as Goliad and Gonzales and near the mouth of the Llano, a branch of the Colorado in N. lat. 30° 20', I was led to the belief that these "villages" were composed by the dirt excavated by these ants, together with refuse vegetable substances gathered by them for food. This also accords with the observation of Mr. Rankin, who is an acute observer and has, in the attempt to exterminate the Ant on his Plantation, destroyed, by digging, many of their "Villages."
>
> These mounds are from a few feet in diameter to fifty yards, and elevated above the common level. They are usually found in the "Post-oak" or sandy soil; and they are sometimes built in the clay soils of the "flats," near watercourses.

> May 1853. Being at Mr. Saml. McFarland's, near Yorktown, Texas, I observed, about twenty five feet below the surface, in a well that was at that time being dug, holes in the sides of the wall about two inches in diameter. These

holes not being perpendicular, were often opened by removing one side of the wall of the hole, leaving its general course and form undisturbed. I asked what caused those holes. The Well-digger, being an old Texian, laughed and called me "green from the States", and said they were ant-holes. I desired them to let me down into the well;—which they did. I there saw, at the depth of twenty five feet, the ants ascending and descending in these holes. It is believed by the old Texians that these ants go to the depth of water. How this may be, I cannot say. I went, however, to an "ant-village" that was about fifty yards distant from the well, and poured two buckets-full of water into the main opening. After about three minutes I saw muddy water, with wet ants and fine sand rolling along down through the aperture in the well. Thus I satisfied myself that they excavated the earth to a great depth.

May 25th. 1853. Walking, this evening, in the valley south of my house I discovered a quantity of newly thrown-up soil, mixed with a soft kind of sand-stone. I found they were brought out by the ants; and knowing there was none of that kind of soil near the surface in that valley, I was led to examine the bank of the creek about one hundred yards distant from the ant-hole; and I there found, at the depth of thirteen feet, the same matters that had been thrown out by those little, very sagacious miners. Curse the little devils!—I wish they were all dead!—for not a cabbage, a plum tree, nor peach tree, and but few ornamental trees can be raised in their vicinity! They attack the largest trees *sans ceremonie*. I saw, three days ago, a very large "China tree" (*Melia azedarach*) hitherto having escaped their ravages, covered with a beautiful foliage. The next morning it was denuded of every leaf—the ground being covered with stems, leaves, and fragments of leaves. Whole armies of ants were engaged in carrying them away about three hundred yards to their "villages." Each one of these depredators, having on his head three short horns, takes hold of a leaf, throws it upon his head, where it is supported by those horns, and walks off with it. Thus he can carry whole grains of Indian corn; and I have seen them marching off with plumseeds. It is amusing to see a company of them with spears of grass, (thrown over their heads, resting between the horns and held by one end in their mandibles,) marching along, in single file, like so many miniature soldiers armed *cap-a-pie*. I saw, but a few days ago, a peach-orchard almost entirely destroyed by these little pillagers. The half-grown fruit was either all

82

cut off, or entirely peeled and left hanging on the trees, and the leaves and soft branches all cut away. "Minutes, New Orleans Academy of Sciences, January 9, 1854."

Walker was an avid amateur natural scientist, and was made a corresponding member of the New Orleans Academy of Sciences. Samuel Wood Geiser, *Naturalists of the Frontier* (Dallas: Southern Methodist University, 1937), p. 335. The first serious student of the Texas leaf-cutting ant, or the agricultural ant as he called it, was Gideon Lincecum, a self-taught natural scientist studying in isolation at Long Point in East Texas. In 1862, a paper Lincecum had prepared on the activities of these ants was read to the Linnaean Society of London by Charles Darwin. Subsequently the society published the paper in its journal. Lois Wood Burkhalter, *Gideon Lincecum, 1793–1874: A Biography* (Austin: University of Texas Press, 1965), p. 213.

25. The reference is to the paper money, actually promissory notes, issued by the Republic of Texas, beginning in 1837. Because of overissue, these notes were quickly plagued by depreciation. A new issue in January, 1839, for example, was valued per dollar at only 37½ cents in specie at the outset. By the winter of 1841–42, these notes had sunk as low as 2 cents on the dollar in some areas. E. T. Miller, "The Money of the Republic of Texas," *Southwestern Historical Quarterly* 52 (1949): 294–300; see also William Gouge, *The Fiscal History of Texas: Embracing an Account of Its Revenues, Debts, and Currency from the Commencement of the Revolution in 1834 to 1851–52* (Philadelphia: Lippincott, Grambo, and Co., 1852); Roemer, *Texas*, p. 88.

26. For a fuller account of the Brazos bottom lands, see Roemer, *Texas*, pp. 74, 75–76.

27. Dry spells and droughts, especially in the western portions of the state, have been a fact of life throughout the course of Texas history. Records do not exist for the one in 1839, but for a general treatment of this subject, see William Curry Holden, *Alkali Trails, or, Social and Economic Movements of the Texas Frontier, 1846–1900* (Dallas: Southwest Press, 1930), chap. 7; see also, Jordan, *Texas: A Geography*, p. 21.

28. Stephen F. Austin's famous map of March, 1830. Some six years in preparation, it is considered the first generally accurate depiction of Texas. James C. Martin and Robert Sidney Martin, *Maps of Texas and the Southwest, 1513–1900* (Albuquerque: Published for the Amon Carter Museum by the University of New Mexico Press, 1984), p. 121.

29. San Bernard River

30. San Felipe, situated on the Brazos River in Austin County, is considered the birthplace of Anglo-American settlement in Texas. Founded in 1823 as San Felipe de Austin, it was the headquarters and unofficial capital of the Austin colony. The first organized opposition to Mexican rule was voiced here in 1832. Three years later, a convention held in San Felipe led to Texas' declaration of independence. San Felipe was burned in 1836 by Sam Houston's retreating army to delay the advance of Santa Anna. The town was rebuilt after the revolution. But with

the removal of the county seat to Bellville in 1846, San Felipe nearly ceased to exist. Dudley G. Wooten, ed., *A Comprehensive History of Texas, 1685–1897* (Dallas: W. G. Scarff, 1898), 1:261–74; *Encyclopedia of Texas*, 2d ed. (St. Clair Stores, Mich.: Somerset Publishers, 1985), pp. 424–25; *The Handbook of Texas* (Austin: Texas State Historical Association, 1952), 2:550; Roemer, *Texas*, p. 77.

31. Roemer, *Texas*, p. 78.

32. Mirages, or optical illusions resulting from the refraction of light through air layers of varying density, were a relatively commonplace natural phenomenon on the expansive Texas prairies. See, for example, Olmsted, *A Journey Through Texas*, pp. 167–69. For an explanation of mirages, see Sybil P. Parker, ed., *McGraw-Hill Dictionary of Scientific and Technical Terms*, 4th ed. (New York: McGraw-Hill Book Company, 1989), p. 1210.

33. Riddell's "great white crane" was apparently a Whooping Crane (*Grus americana*). Now extremely endangered, this largest of America's two native crane species was indigenous to the coastal prairies of Texas during the early nineteenth century. Harry C. Oberholser, *The Bird Life of Texas* (Austin: University of Texas Press, 1974), 1:286–88; see also, Robert Porter Allen, *The Whooping Crane* (New York: National Audubon Society, 1952).

34. While Riddell attributed their presence to flood waters, these pebbles, called cobbles, were from the lower sandstone bed of the Cenozoic formations that made up this region and exposed through the forces of erosion. E. H. Sellards, W. S. Adkins, and F. B. Plummer, *The Geology of Texas*, volume 1: *Stratigraphy*, University of Texas Bulletin no. 3332 (Austin, 1932), pp. 753, 758–59. Primitive stones, according to Amos Eaton's geological nomenclature employed by Riddell, were "those which contain no organic relics nor coal." Amos Eaton, *A Geological Nomenclature For North America; Founded Upon Geological Surveys, Taken Under the Direction of The Hon. Stephen Van Rensselaer* (Albany: Packard and Van Benthuysen, 1828), p. 3.

35. Flowering Spurge

36. (*Chamaesyce hypericifolia*); Larger Spotted Spurge

37. A genus of herbs, shrubs, and trees.

38. (*Eryngium yuccifolium*); Rattlesnake-Master

39. Hospitality in early Texas had its limitations. Sibley, *Travelers in Texas*, pp. 38–39

40. His malady was malaria, a leading disorder in this unhealthy region. Pat Ireland Nixon, the most prominent historian of medicine in early Texas, noted: "There were many factors which retarded development of the Texas Republic. Not the least of these factors was disease. And of these diseases, malaria was most important." Pat Ireland Nixon, *The Medical Story of Early Texas, 1528–1853* (Lancaster, Penn.: Mollie Bennett Lupe Memorial Fund, 1946), p. 286. Because of a shortage of physicians and the ineffectiveness of the medical profession in combating the era's principals diseases, numerous early

Texans practiced self-dosage, depending on botanical remedies prepared from indigenous plants. Substitutes for quinine, the specific for malaria, included a wide variety of bitter tasting plants: ague tree, wild basil, squaw mint, deer tongue, boneset, gall flower, snake root, flowering dogwood, and golden rod. Ibid., pp. 311–83.

41. Riddell first saw the Colorado River in the Gulf Coastal Plain physiographic province of Texas, the youngest of the state's geologic provinces. Once a portion of the floor of the Gulf of Mexico, this vast region ranges in age from 2 million to 138 million years, its surface materials dating from the Pleistocene Epoch and the Tertiary and Cretaceous Periods. The area was largely formed by marine and continental deposits. One product of these processes is sandstone. In Riddell's usage, amorphous most likely meant nondescript in structure. Sellards and Baker, *Geology of Texas*, pp. 33–48; *Handbook of Texas*, I, p. 680; Jordan, *Texas: A Geography*, p. 7.

42. Henry Wax Karnes was a distinguished veteran of the Texas Revolution, for whom Karnes County in South Central Texas is named. In December, 1838, he was authorized to raise troops to war against the Comanches. It was not until June, 1839, however, that he issued a call for volunteers. At the time of Riddell's trip, he was assigned to protect surveyors running the line between Bexar and Bastrop counties. Since this activity was in the area of the lost San Saba mine, Riddell and his associates apparently decided to join Karnes's force for protection from the Comanches. But as subsequent events would show, Karnes, too, had an interest in the lost treasure. In fact, guarding the surveyors may have been a pretext for him to search for it. Sam Houston Dixon and Louis Wiltz Kemp, *The Heroes of San Jacinto* (Houston: Anson Jones Press, 1932), pp. 307–308.

43. There is no George W. Reese listed in White, ed., *1840 Census;* the sole Joseph Hopkins was a resident of Harris County (p. 66); John S. McDonald was probably John McDonald of Bexar County (p. 15).

44. Columbus, the county seat of Colorado County, was founded in 1823 on the site of an old Indian village by members of the Austin colony. *Encyclopedia of Texas*, p. 257; *Handbook of Texas*, 1:382; Roemer, *Texas*, p. 81.

45. Riddell was in the proximity of the dividing line between the Post Oak Belt and the Blackland Prairie region. The former, lying to the west of the Coastal Prairie zone, was characterized by post oak, hickory, and other oaks in conjunction with numerous scattered prairies, while the latter, which separated the Post Oak Belt and the Eastern Cross Timbers and is considered the outstanding prairie in Texas, consisted of lush, tall grasses in the uplands and hardwoods along water courses. Jordan, *Texas: A Geography*, pp. 30–31.

46. Navidad River

47. White Water Lily

48. (*Brasenia schreberi*)

49. Part of the algae genus.

50. The flight of Texans from the invading army of Santa Anna beginning in January 1836 is known as the Runaway Scrape. Kate Scurry Terrill, "The 'Runaway Scrape,'" in Wooten, ed., *Comprehensive History of Texas*, vol.1, chap. 13.

51. Lavaca River

52. This is the Nuttall nomenclature for the prickly pear genus. The modern one is *Opuntia*.

53. Roemer, *Texas*, p. 116.

54. Both the Eastern Turkey (*Meleagris gallopavo silvestris*) and the Rio Grande Turkey (*Meleagris gallopavo intermedia*) were indigenous to Texas. Riddell probably saw the latter, similar to but slightly smaller than the more geographically widespread former variety. The dividing line between the two was generally thought to be the junction of the East Texas Timber Country and the Blackland Prairie, or roughly the line of the Brazos River. Arlie William Schorger, *The Wild Turkey: Its History and Domestication* (Norman: University of Oklahoma Press, 1966), chap. 3.

55. The *Handbook of Texas* (2:495–96) points out that at least twenty Texas streams have this name. None of those listed, however, seem to be the one to which Riddell is referring.

56. Probably B. D. McClure. White, ed., *1840 Census*, p. 58.

57. On the Comanches and their activities in Texas, see Ernest Wallace and E. Adamson Hoebel, *The Comanches: Lords of the South Plains* (Norman: University of Oklahoma Press, 1952) and Rupert Norval Richardson, *The Comanche Barrier to South Plains Settlement* (Glendale, Calif.: A. H. Clark Company, 1933); Roemer, *Texas*, pp. 274–81.

58. This Riddell's phonetic spelling of *cavallard*, a corruption of the Spanish word *caballada* (a drove of horses). *Texas in 1837*, p. 206 (note 4).

59. A Spanish settlement dating from 1749, and one of the oldest municipalities in Texas. In 1829, its name was changed to Goliad. *Encyclopedia of Texas*, pp. 307–308; *Handbook of Texas*, 1:699, 2:1–2.

60. The horsemanship of the Comanches was legendary, contributing to their reputation as "the finest light cavalry the world has ever seen." Quoted in John K. Herr and Edward S. Wallace, *The Story of the U.S. Cavalry, 1775–1942* (Boston: Little, Brown, 1953), p. 69.

61. This stream is not among the four in Texas that are known as Black Creek. *Handbook of Texas*, 1:168–69.

62. Thomas Drummond was an eminent Scottish naturalist who collected in Texas during 1833 and 1834, chiefly in the Austin Colony and on Galveston Island. Geiser, *Naturalists of the Frontier*, pp. 73–105; Susan Delano McKelvey, *Botanical Exploration of the Trans-Mississippi West, 1790–1850* (Jamaica Plain, Mass.: Arnold Arboretum of Harvard University, 1955), pp. 486–507.

63. Mesquite. The mesquite tree is a member of the *Prosopis* genus of the Leguminosae Family, a very large family of herbaceous or woody plants. It was not indigenous to the grassland prairies of Texas.

Correll and Johnston, *Manual of the Vascular Plants of Texas*, pp. 783–84; Jordan, *Texas: A Geography*, p. 31.

64. The seat of Gonzales County, Gonzales was settled in 1825 and named for Don Rafael Gonzales, the provincial governor of the Mexican province of Coahila and Texas. Because of its location on the western boundary of Anglo-American Texas, it was the scene of considerable Texas revolution activity and has been called the "Lexington of Texas." On March 12, 1836, Gonzales was burned in Houston's retreat after the fall of the Alamo. The town was rebuilt after the battle of San Jacinto. *Encyclopedia of Texas*, pp. 308–309; *Handbook of Texas*, 1:707–708; Roemer, *Texas*, pp. 86, 293.

65. This was apparently Reuben Ross. A veteran of the Texas Revolution, Ross was at this time captain of the Gonzales company of frontier rangers. He was on his way to the Rio Grande to become involved as a soldier of fortune in one of the numerous revolts and counterrevolts that wracked the young Mexican republic. The principal antagonists were the Centralists, who were the proponents of a strongly centralized government, and the Federalists, who sought a confederation and the limiting of the functions of the national government to a bare minimum. *Handbook of Texas*, 2:507; Michael C. Meyer and William L. Sherman, *The Course of Mexican History*, 4th ed. (New York: Oxford University Press, 1991), pt. V.

66. The danger, as the foregoing suggests, was posed by the Comanches. "The Indians," a leading citizen of San Antonio asserted, "were always lurking around in small bodies hiding close to town, waiting for an opportunity to strike without danger to themselves." Mary A. Maverick and George Madison Maverick, *Memoirs of Mary A. Maverick*, ed. Rena Maverick Green (San Antonio: Alamo Printing Company, 1921), p. 50.

67. Probably John G. King. See White, ed., *1840 Census*, p. 56.

68. Seguin was a new town at the time of Riddell's visit. The first permanent settlement at this crossing of the Guadalupe River was not established until 1838. Initially called Walnut Springs, the name was changed to Seguin in 1839 in honor of Col. Juan N. Seguin, the commander of the only detachment of Texas-born Mexicans in the Battle of San Jacinto. *Encyclopedia of Texas*, p. 430; *Handbook of Texas*, 2:590–91; Roemer, *Texas*, p. 90.

69. A fountain is a spring of water issuing from the earth. It is synonymous with fount. Robert L. Bates and Julia A. Jackson, eds., *Glossary of Geology*, 3rd ed. (Arlington, Va.: American Geological Institute, 1987), p. 257.

70. Known for its distinctive bluish waters, the natural beauty of the Guadalupe was generally agreed upon by travelers. Francis Moore, Jr., *Map and Description of Texas, Containing Sketches of Its History, Geology, Geography and Statistics* (Philadelphia: H. Tanner, Junr., 1840), pp. 76–77; Roemer, *Texas*, p. 114.

71. The mesquite was of considerable interest to travelers. *Texas in 1837*, pp. 89–90; Roemer, *Texas*, pp. 114–15.

72. Riddell surmised incorrectly in this instance. Mesquite grass belongs to the *Brachiaria* genus.

73. Roemer, *Texas*, pp. 117, 151.

74. Although its origin is unclear, "John" seems to have been a term of derision that was applied to minorities of various types. David R. Goldfield, *Promised Land: The South Since 1945* (Arlington Heights, Ill.: Harlan Davidson, 1987), p. 107.

75. A native of North Carolina, Lindsay S. Hagler was admitted to the United States Military Academy in 1833. He left in 1836, seemingly to participate in the Texas Revolution. He served from June, 1836 to December, 1837. In 1839, Hagler offered his services to the Mexican Federalist forces. He later served in the Congress of the Republic of Texas. *Register of the Graduates and Former Cadets, United States Military Academy 1802–1964* (New York: West Point Alumni Foundation, 1964), pp. 223–24; Texas House of Representatives, *Biographical Directory of the Texan Conventions and Congresses, 1832–1845* (Austin, 1941), p. 93.

76. (*Cladrastis lutea*; synonymous with *Cladrastis kentukea*); Riddell's uncertainty was well-founded, as this genus is not indigenous to Texas. He subsequently refers to this tree as the Wild China. Apparently, he confused it with the Soapberry tree (*Sapindus drummondii*), which was also commonly called the Wild China or Wild Chinaberry. Frank W. Gould, *Texas Plants: Checklist and Ecological Summary*, rev. ed. (College Station: Texas Agricultural Experiment Station, 1975), p. 59.

77. (*Quercus virginiana*); Live Oak

78. Hackberry

79. (*Taxodium distichum*); Bald Cypress

80. Common Balloon Vine

81. Although Roemer also reported seeing the *Yucca filamentosa* in this area, these two specimens are not among the twenty types of yucca considered indigenous to Texas. Correll and Johnston, *Manual of the Vascular Plants of Texas*, pp. 395–402.

82. Salado Creek, a stream on the Old San Antonio Road.

83. Riddell was leaving the Gulf Coastal Plain and entering the Balcones fault zone. Jordan, *Texas: A Geography*, pp. 7–8; Roemer, *Texas*, pp. 117–18.

84. Mustangs were the descendants of Spanish horses brought to the American Southwest by the conquistadors. They are celebrated in an extensive literature. J. Frank Dobie, Mody C. Boatright, and Harry H. Ransom, eds., *Mustangs and Cow Horses* (Austin: Texas Folklore Society, 1940); Don Worcester, *The Spanish Mustang: From the Plains of Andalusia to the Prairies of Texas* (El Paso: Texas Western Press, 1986).

85. 2 Chron. 8:4; a city built by Solomon in the Syrian desert to facilitate trade with the East. It was located in a fertile oasis 140 miles northeast of Damascus. Paul J. Achtemeier, et al., *Harper's Bible Dictionary* (San Francisco: Harper & Row, 1985), p. 1015. Such a reaction to San Antonio on first sight was nearly universal See, for example,

*Texas in 1837*, pp. 94–95; Roemer, *Texas*, pp. 133–34. According to Sibley, San Antonio was "the most interesting town to travelers between 1761 and 1860." Sibley, *Travelers in Texas*, pp. 61–63.

86. Sunflower family

87. The "verdant appearance" of the San Antonio valley was the result of a system of irrigation using the waters of the San Antonio River. This irrigation system, Sibley remarks, was a source of frequent admiration by early travelers. Sibley, *Travelers in Texas*, p. 52.

88. The best recent work on "the cradle of Texas independence" is Susan Prendergast Schoelwer, with Tom W. Glaser, *Alamo Images: Changing Perceptions of a Texas Experience* (Dallas: DeGolyer Library and Southern Methodist University Press, 1985); see also, Roemer, *Texas*, pp. 124–25.

89. Although a metropolis of nearly a million today, San Antonio, owing to devastations and dislocations of first the Mexican and then the Texas war for independence, was a city in decline at the time of Riddell's visit. Its population had fallen from 5,000 in 1810 to 2,400 in 1834 and to 1,800 in 1839. It would dip to 600 in 1843. Ray F. Broussard, "San Antonio During the Texas Republic: A City in Transition," *Southwestern Studies* 18 (1967): 5, 29; John L. Davis, *San Antonio: A Historical Portrait* (Austin: Encino Press, 1978); Roemer, *Texas*, p. 120.

90. Narrow-Leaved Cat-Tail

91. Called *jacales*, these crude structures, often consisting of a single twelve-foot by twelve-foot room, persisted as the residences of Tejanos into the early twentieth century. Arnoldo De León, *The Tejano Community, 1836–1900* (Albuquerque: University of New Mexico Press, 1982), pp. 114–21. For the persistence of the *jacal* in San Antonio, see Davis, *San Antonio*, p. 78. A good overview of this type of structure is Willard B. Robinson, "Colonial Ranch Architecture in Spanish-Mexican Tradition," *Southwestern Historical Quarterly*, 83 (1979):123–50. While chimneys built of sticks and plastered with a thick layer of mud or clay allowed indoor baking, most food was prepared in semidome structures of baked and blackened mud situated some twenty or thirty feet from the house. De León, *The Tejano Community*, pp. 117–18.

92. Tejano dress, according to De León, was a function of circumstance and occasion. De León, *The Tejano Community*, pp. 133–35; see also, Roemer, *Texas*, p. 124.

93. Town square

94. White, ed., *1840 Census*, p. 13. Egery was one of the leaders in the scheme to locate the lost San Saba mine.

95. This is probably William Hester Patton. Patton emigrated to Texas from Kentucky in 1828. He served with the Texan forces throughout the Texas Revolution. He participated in the siege of Bexar, was an aide to Sam Houston at the battle of San Jacinto, and was later appointed quartermaster general. After the war, he represented Bexar County in the Second Congress of the Republic. A surveyor by profession, Patton settled on the San Antonio River some thirty-five miles below San Antonio. He was murdered in June, 1842, by a band of

Mexicans. Dixon and Kemp, *Heroes of San Jacinto*, pp. 46–47; *Handbook of Texas*, 2:346; White, ed., *1840 Census*, p. 16.

96. Probably H. A. Alsbury. White, ed., *1840 Census*, p. 12.

97. Dancing, as the great popularity of the fandango illustrates, was a favorite form of entertainment for the citizens of San Antonio. Broussard, "San Antonio During the Texas Republic," p. 22; Roemer, *Texas*, pp. 121–24, 131–33.

98. The culprit seems to have been the Comanche. "Indian braves," Sibley wrote, "measured their manliness by the number of horses they had stolen...." "All who came in close contact with the Indians," she added, "complained of their continual thievery." Sibley, *Travelers in Texas*, p. 29 (first quotation), p. 72 (second quotation). For a more detailed account of the horse-stealing activities of Indians in San Antonio, see *Texas in 1837*, pp. 110–11. Roemer, who noted the theft of horses by the Comanches, also reported that it was widely believed among settlers "that the Mexicans have the same inclination and skill for stealing horses as the Indians." Roemer, *Texas*, pp. 158, 275–76.

99. On the many uses of rawhide, see Raines and Barnes, *Cattle*, Appendix 1.

100. Tufa is spongy, porous rock composed of calcium carbonate. Robert Scott Mitchell, *Dictionary of Rocks* (New York: Van Nostrans Reinhold, 1985), 202. The missions, Sibley observed, were "by far the greatest curiosities in the neighborhood of San Antonio." Sibley, *Travelers in Texas*, p. 63.

101. Saint Louis Cathedral

102. Five Spanish missions were located in San Antonio and its environs: San Antonio de Valero (the Alamo), San José de Aguayo, San Juan Capistrano, San Francisco de la Espada, and Nuestra Señora de la Purísima Concepción. As Riddell earlier noted in his diary, his destination was San José de Aguayo, the so-called "queen of the missions," situated 7½ miles down the San Antonio River from the Alamo. Roemer, *Texas*, pp. 126–30; James Day et al., *Six Missions of Texas* (Waco: Texian Press, 1965), pp. 129–64; Edward W. Heusinger, *Early Explorations and Mission Establishments in Texas* (San Antonio: Naylor Company, 1936), chaps. 5–6, 8.

103. Olmos Creek. *Handbook of Texas*, I, p. 313.

104. Bastrop County, created in 1836, was one of the twenty-three original counties of the Republic of Texas. It was named for Baron Felipe Enrique Neri de Bastrop who was honored for helping Moses Austin secure land from the Mexican government for the settlement of Americans in Texas and for his aid to Stephen F. Austin as the Texas revolution neared. At this time Bastrop and Bexar counties shared a common boundary. *Handbook of Texas*, 1:120–21; Kenneth Kesselus, *History of Bastrop County, Texas, Before Statehood* (Austin: Jenkins Publishing Company, 1986), chaps. 9–10. On the rigors and dangers of surveying in frontier Texas, see Roemer, *Texas*, p. 224.

105. This Edward H. Barton, Riddell's colleague in the Medical Department of the University of Louisiana and antagonist in the controversy over the etiology of yellow fever.

106. White, ed., *1840 Census*, p. 16. Riddell's name was variously spelled; see Thomas Cary Johnson, *Scientific Interests in the Old South* (New York: D. Appleton-Century Company, 1936), pp. 48–49.

107. Riddell was traveling across the Edwards Plateau, a subdivision of the Great Plains. This gently rolling to rugged, semiarid physiographic province lies to the north and west of the Balcones escarpment and is composed of durable Lower Cretaceous limestone. The Cretaceous is one of the most significant geologic formations in Texas, containing some of the state's best water resources, richest soils, and largest oil deposits. Delphinula are ocean-dwelling snails of the superfamily Trochacea. The shells of this superfamily are turban-shaped, globular, or smoothly conical, with no slit or sinus. The inner shell layer is pearly. Sellards, Adkins, and Plummer, *Geology of Texas: Stratigraphy*, pp. 259– 65; Sellards and Baker, *Geology of Texas: Structural and Economic Geology*, p. 28; Jordan, *Texas: A Geography*, p. 12; Ida Thompson, *The Audubon Society Field Guide to North American Fossils* (New York: Knopf, 1982), p. 416.

108. Such a state of friendliness between Mexicans and Indians does not appear to have been the usual situation. Rather, there seems to have been a pronounced mutual antipathy. Mary Austin Holley, for example, strongly stressed this point, writing: "If a hunting party takes the life of a North American after making him prisoner, without bringing him before the council for trial, the offenders are punished with death. Not so with the Mexicans, who are considered as enemies and treated as such. This hatred is mutual, and fully reciprocated on the part of the Mexicans. Hence the origin of the epithet expressing odium, so general in all parts of Mexico: to denote the greatest degree of degradation, they call a person a Comanche." Mary Austin Holley, *Texas* (Lexington: J. Clarke, 1836; reprint, Austin: Steck Company, 1985), pp. 54–55.

109. Although skeptical of its existence, Roemer also searched for this fabled mine while on an exploring expedition to the region in February, 1847. Roemer, *Texas*, pp. 256–59.

110. Holley, *Texas*, pp. 161–73.

111. Riddell was entering that portion of the Edwards Plateau commonly called the Hill Country of Central Texas because of its distinctive sculpturing. The native plant life of this region consists of scattered stands of live oak, shinnery oak, red oak, and juniper. Jordan, *Texas: A Geography*, pp. 12, 29. Roemer, who spent most of his stay in Texas in New Braunfels, has much to say about this region. Roemer, *Texas*, chs. 7–24, passim.

112. (*Ilex decidua*); Deciduous Yaupon Holly

113. Ironwood genus

114. Indian-Cherry

115. (*Cocculus carolinus*); Red-Berried Moonseed, Snailseed

116. (*Parthenocissus quinquefolia*); Virginia Creeper

117. Heartleaf

118. Bears were in abundance during the early years of colonization in Texas; by the mid-1830s, they were in retreat; and at mid-century, they were rapidly disappearing. Jordan, *Texas: A Geography*, pp. 43–44. Viktor Bracht, a German settler, identified three kinds of bears in Texas in 1848: the grizzly, the black bear, and a slender, fleet species that Americans called "racer." Viktor Bracht, *Texas in 1848*, trans. Charles Frank Schmidt (San Antonio: Naylor Printing Company, 1931), p. 43.

119. Now Sabina Creek

120. A silicious rock is one containing a high percentage of silica. A buhrstone is a silicious rock suitable for use as a millstone. Jordan, *Texas: A Geography*, p. 12; Mitchell, *Dictionary of Rocks*, p. 36.

121. Sumac genus

122. Venus' Hair Fern

123. (*Asplenium platyneuron*); Ebony Spleenwort

124. A fern seemingly named by Riddell, but apparently not recognized by either contemporary or succeeding taxonomists. J. L. Riddell, "Catalogus Florae Ludovicianae," *New Orleans Medical and Surgical Journal* 8 (1852): 743–64.

125. Riddell was accustomed to seeing the cypresses of southern Louisiana with their infestations of Spanish Moss. This commensal (*Tillandsia usneoides*) was largely limited to the swamps of the Gulf Coast. *Gray's Manual of Botany*, 392.

126. Bees were plentiful in early Texas. Roemer, *Texas*, p. 265; Holley, *Texas*, p. 68; *Texas in 1837*, p. 92; Bracht, *Texas in 1848*, pp. 50–51.

127. The dark loamy mollisols soil order predominates on the Edwards Plateau. Jordan, *Texas: A Geography*, p. 37.

128. The Cretaceous limestones contain a variety of marine fossils. Sellards, Adkins, and Plummer, *Geology of Texas: Stratigraphy*, p. 268. Favosites was a Paleozoic coral. Richard Owen, *Palaeontology, or A Systematic Summary of Extinct Animals and Their Geological Relations* (Edinburgh: A. and C. Black, 1860), p. 23.

129. Again Riddell's habit of spelling phonetically led to misspellings: panola should be pinóle and matat should be metate. De León, *Tejano Community*, p. 122.

130. Metating.

131. Winged-elm

132. Red Oak

133. (*Quercus falcata*); Spanish Oak, Southern Red Oak

134. Pin Oak

135. Flowering Dogwood

136. (*Carya illinoensis*); Pecan

137. Dwarf Sumac, Shining Sumac, Wing-Rib Sumac

138. Hop Tree

139. Eastern Red Cedar

140. Fragrant Sumac

141. *Vitis cordata* is synonymous with *Ampelopsis cordata;* Heartleaf.

142. These were the fabled Texas longhorns. J. Frank Dobie, *The Longhorns* (Boston: Little, Brown and Company, 1941).

143. In Texas the natural habitat of the buffalo, actually the bison (*Bison bison*), lay north and west of a line roughly circumscribed by the 31st degree of latitude and the 98th meridian. Riddell's party was approximately forty miles below this line. Francis Haines, *The Buffalo* (New York: Crowell, 1970), pp. 8–9; Larry Barsness, *Heads, Hides & Horns: The Compleat Buffalo Book* (Fort Worth: Texas Christian University Press, 1985); Roemer, *Texas*, pp. 192, 198–200.

144. Intrigued by this natural phenomenon, Riddell published his views on it after returning from Texas. "While in Texas the second time," he wrote,

> I had full opportunity to study the phenomenon of "hog-wallow prairies." The long droughts in summer cause the woodless surface of the prairies to crack deeply, and oftentimes symmetrically; subsequent rains wash the adjacent earth into these cracks. filling them up, converting them into little valleys, and leaving intermediate hillocks. Next year the same round of cause and effects occurs in the same places, and thus successive years contribute for a long time to produce a maximum effect, the appearance of which is very striking. When the prairie is level, the hillocks are exactly hexagonal, and usually eight or ten feet in diameter. The depressions between them are commonly twelve or eighteen inches deep. If the surface is inclined, the hexagons become elongated at right angles to the direction of the dip, when they frequently resemble the waves of the ocean. From difference of surface, soil, and exposure, there arises a great diversity in the size, depth, and general appearance of the hog-wallows. They never occur in a sandy soil, consequently they are not seen on the sandy prairies near the sea coast. I do not remember to have seen them among the mountains in the Comanche country; but else they frequently greet the eye of the traveller in most every part of Texas.

J. L. Riddell, "Hog Wallow Prairies," *American Journal of Science and Arts* 39 (1840): 211–12.

Roemer also observed this phenomena, describing it as "a prairie whose surface is covered with countless flat, irregular depressions about five to six feet in diameter." But unlike Riddell, he offered no view on its cause. Roemer, *Texas*, pp. 183–84. A hog wallow, according to a modern reference work on geology is "a faintly rolling land surface characterized by many low, coalescent or rounded mounds . . . that are slightly higher than the basin-shaped depressions between them." These depressions, it is believed, are formed by heavy rains. Bates and Jackson, *Glossary of Geology*, pp. 293–94.

145. See note 76.

146. While the identity of Fred's malady is unknown, contemporary accounts point out that fevers were a common source of morbidity. Nixon, *Medical History of Early Texas*, p. 284. The expedition's physician was seemingly Dr. Weideman of San Antonio. Weideman, a colorful figure, figured memorably in the so-called Court House Massacre of March, 1840, a bloody confrontation between a group of Comanches who had come into San Antonio to treat for peace and to exchange captives and citizens of the city. "Connected with the battle in a rather macabre way," a biographer of San Antonio has written,

> was a former Russian, Dr. Weideman. Sent by the Emperor of Russia to report on Texas, he had stayed. At the time of the Indian fight he was a 35-year-old practicing physician who lived on the acequia just north of the Main Plaza. He often rode as a Ranger and cultivated a host of strange interests.
>
> On the evening in question, Mrs. [Mary A.] Maverick and Mrs. Higginbotham were talking of the battle. Dr. Weideman suddenly appeared outside the window that fronted the street. With a courteous, "with your permission, Madam," he placed the decapitated head of an Indian on the window sill for safekeeping.
>
> Soon he was back with another. He explained to the ladies that the heads, male and female, were for his studies—he wanted the skulls. In a rush of scientific zeal, he departed to his house, where he boiled the flesh from the bones.
>
> During the night of March 20 he dumped the remains—less the skulls—into the main acequia. Now, while the two rivers of the city were clearly for washing, bathing, and even a little dumping, the acequias were not. The two artificial channels running through the settlement from near the river's headwaters were kept pure for drinking, quite distinct from the rivers. The acequia which ran by the doctor's house passed San Fernando Church on its sparkling way, before entering the residential area to the south.
>
> The next day the doctor was found out and confronted at the mayor's office by loudly exclaiming men and shrieking, vomiting women. "Diabolo," they called him. "Demonio," they cried, with a sprinkling of the perhaps more temperate "sin verguenza." All of which he took calmly, paid his fine, and went off chuckling.
>
> The doctor was indeed a strange person, fond of pet snakes and given to proving theft through a formal ceremony. He would appear dressed as a sorcerer, strange-figured robe and all, behind a boiling cauldron. The liquid, when cooled, had the supposed property of changing any thief's hand black when dipped therein—others being left untainted.

The Mexicans of San Antonio—those who crossed themselves and avoided him on the streets—said the doctor was in league with the devil. He may indeed have been: it was also said that he did not take pay for his medical services.

Davis, *San Antonio*, pp. 15–16; see also, Maverick and Maverick, *Memoirs of Mary A. Maverick*, pp. 38–41

147. Although wolves can vary greatly in color, there is no recognized species of white or albino wolf. The principal species in early Texas were the Texas Gray Wolf (*Canis lupus monstrabilis*) and the Texas Red Wolf (*Canis niger rufus*). The former was the larger and more numerous of the two. It was especially prevalent in the western portion of the states. The Texas Gray Wolf was usually dark in color but could vary to nearly white. It was probably the San Saba white wolf in question. The description of a white wolf slain in this same general region in 1828 by Jean Louis Berlandier, the eminent European naturalist who traveled in Mexico between 1826 and 1834, seems to verify this observation. Edward A. Goldman, *Classification of Wolves*, Part II of *The Wolves of North America* (Washington, D.C.: American Wildlife Institute, 1944), pp. 49, 466–68, 486–89; David E. Brown, ed., *The Wolf in the Southwest: The Making of an Endangered Species* (Tucson: University of Arizona Press, 1983); Jean Louis Berlandier, *Journey to Mexico During the Years 1826 to 1834*, trans. Sheila M. Ohlendorf, Josette M. Bigelow, and Mary M. Standifer (Austin: Texas State Historical Association, 1980), 2:350–51. Travelers in Texas routinely referred to the presence, varieties, and behavior of the wolf. Roemer, *Texas*, pp. 80, 84, 113, 233.

148. As Riddell speculated, ranching became the principal economic activity in the southern Edwards Plateau. Jordan, *Texas: A Geography*, chap. 8.

149. Buttonbush

150. The cry of the Sandhill Crane (*Grus mexicana*), an authoritative guide on American birds points out, "is a veritable voice of Nature, untamed and unterrified. Its uncanny quality is like that of the Loon, but is more pronounced because of the much greater volume of the Crane's voice. Its resonance is remarkable and its carrying power is increased by a distinct tremolo effect. Often for several minutes after the birds have vanished, the unearthly sound drifts back to the listener, like a taunting trumphet from the under-world." The blue crane referred to by Riddell was likely not a crane at all but the Great Blue Heron. T. Gilbert Pearson, ed., *Birds of America* (Garden City, N.Y.: Garden City Books, 1936), pp. 200–201; see also, Oberholser, *Bird Life of Texas*, 1:288–89.

151. Pedernales River

152. Riddell had entered the southern sector of the Llano Basin. Lying at the geographic center of the state, this area has been called one of the most interesting physiographic sections in Texas. In reality,

it is a 1,000-foot-deep erosional basin that was formed by the Colorado River. The Precambrian formation that makes up the floor of this geologic structure represents the most extensive exposure of the oldest rock in Texas. Sellards and Baker, *Geology of Texas: Structural and Economic Geology*, pp. 73–85; Jordan, *Texas: A Geography*, p. 12; *Handbook of Texas*, I, p. 680; Roemer, *Texas*, chaps. 19–23.

153. Groundsel genus. This was a new variety and it was named *Senecio Riddellii* by Gray and Torrey in Riddell's honor. Torrey and Gray, *A Flora of North America*, 2:444.

154. A not infrequently found but difficult to document reference in the historical literature of Texas and the Southwest. *Handbook of Texas*, 1:386. Riddell's favorable impression of the Llano Estacado was not generally shared by other travelers who dreaded crossing this massive geographical obstacle. Sibley, *Travelers in Texas*, p. 53.

155. Riddell is using spy in the context of to spy out.

156. Lucifer is synonymous with friction match. William S. Walsh, *A Handy Book of Curious Information* (Detroit: Gale Research Company, 1970), pp. 527–31.

157. It was not uncommon for the Comanches to adopt captured boys as brothers; girls were often wed by their captors. Wallace and Hoebel, *The Comanches*, p. 241; Sibley, *Travelers in Texas*, pp. 80–82; Roemer, *Texas*, pp. 242–43, 271.

158. Saliferous rock, according to Eaton, belonged to the secondary class of rocks, or "those which contain, in some localities, dryland or fresh-water organic relics, as well as marine, or bituminous coal." This type of rock consisted of "red, or bluish-grey, sand or clay-marle, or both." Eaton, *Geological Nomenclature*, pp. 3, 13.

159. Although it is difficult, if not impossible, to be sure, internal evidence in the subsequent text suggests that this stream is Sandy Creek, rising in northeastern Gillespie County and flowing into the Colorado in southeastern Llano County.

160. In Eaton's nomenclature primitive rocks, as previously noted, were those that contained no organic remains or coal. Granular quartz, he explained, consisted of "grains of quartz united without cement." Eaton, *Geographical Nomenclature*, pp. 3, 11.

161. For unclear reasons, perhaps because of competition from Karnes's party, Riddell and his associates have abandoned their attempt to locate the lost San Saba mine at this point.

162. White, ed., *1840 Census*, pp. 15, 18.

163. Ibid., p. 14. A Mississippian who came to Texas as a surveyor, Hays commanded a company of rangers paid first by the Republic of Texas and then by the United States to help protect western Texas. Roemer, *Texas*, p. 131.

164. White, ed., *1840 Census*, p. 14.

165. Ibid., p. 17.

166. Ibid., pp. 16, 18.

167. Ibid., p. 17.

168. Ibid., p. 14.

169. A Mexican town located just below the Rio Grande on the Presidio Road that ran southwest from San Antonio. See Arrowsmith, "Map of Texas," in Martin and Martin, *Maps of Texas and the Southwest,* p. 126.

170. "Old Gonzales" may have been one of several people. First, he may have a local strongman. Second, Riddell may have been referring to Rafael Gonzales, the Mexican military figure for whom the town of Gonzales, Texas, was named. In 1834, he was appointed secretary of the Commandancia of Coahuilla, where he may have become involved in the revolutionary occurrences of 1839 that pitted the Centralists against the Federalists. See, Wooten, *A Comprehensive History of Texas,* 1:126; Vito Alessio-Robles, *Coahuilla y Texas, Desde la Consumación de la Independencia Hasta el Tratado de Paz de Guadalupe Hidalgo* (Mexico, 1945–46), 1:194. Third, Riddell may have been describing Col. José Maria Gonzales. A member of a prominent Mexican family, Gonzales supported Texas' break with Mexico in 1835. He subsequently joined the Texas forces and fought with distinction. After the war, he apparently returned to Mexico, settling near the Rio Grande. A Centralist, he and his supporters were forced to flee to the United States by the Federalist forces in 1839. See Wooten, *A Comprehensive History of Texas,* I, pp. 205, 349.

171. Enchanted Rock is "the second largest elevated granite batholith in America." (Stone Mountain in Georgia is the largest.) Situated in eastern Llano County near the Gillespie County line, it towers 325 feet above Sandy Creek and covers more than a square mile in the basin between the Llano and Pedernales rivers. The prominence gets its name from Indian legends, which hold that it has a spirit. Charles Rossman, with David Stark and Pat Brown, *Enchanted Rock: Views of a Texas Batholith* (Austin: Duncan & Gladstone, 1985), p. 3; *Handbook of Texas,* pp. 566–67.

172. The Llano Basin, as mentioned, contains the oldest rock materials in Texas, dating from the Precambrian Period. "These ancient or archean materials," a pioneer Texas geologist has written, "consist of granites, gneisses, schists, and marbles, which are cut by numerous intrusions of eruptive rocks, and are highly altered by metamorphism." E. T. Dumble, "Physical Geography, Geology, and Resources," in Wooten, ed., *A Comprehensive History of Texas,* 1:472.

173. Formations of quartz are abundant in the Precambrian rocks of the Llano uplift. Sellards and Baker, *Geology of Texas: Structural and Economic Geology,* p. 270.

174. The quality of the Comanche horse, largely the mustang, has been greatly overrated. "Most of the horses, as a rule," Roemer wrote, "are unsightly and small." Roemer, *Texas,* p. 268.

175. Riddell was apparently correct in his opinion that this river was the Colorado. The western boundary of Bastrop County in 1839 crossed the Colorado in what is today southeastern Llano County and

southwestern Burnet County. Kesselus, *Bastrop County Before Statehood*, p. 193.

176. This scene is perhaps indicative of the wolf's not infrequent tendency to kill all out of proportion to his need for food. Barry Holstun Lopez, *Of Wolves and Men* (New York: Scribner, 1978), chap. 3.

177. (*Populus deltoides*)

# "Geology of the Trinity Country, Texas"

Art. II—*Observations on the Geology of the Trinity Country, Texas, made during an excursion there in April and May, 1839*. By J. L. Riddell, M.D., Professor of Chemistry in the Medical College of Louisiana.

It is well known that, as you proceed inland from the Gulf of Mexico in Eastern Texas, to the distance of eighty or ninety miles, the face of the country presents a general plain, almost as level as the surface of the ocean. As it is elevated thirty or forty feet above tide water, it is necessarily furrowed by water courses; but its most remarkable feature where prairies prevail is the existence of multitudes of wet places, each covering from a few rods to an acre or two in extent, and having a depression of one or two feet below the general level; while always around the margins of these low places are several rounded mounds, having a base of ten or twelve feet, and a height perhaps one fourth as great.

Compared with the age of the main American continent, all this land may be considered as having quite recently emerged from the dominion of the sea. It is essentially a vast deposit of sea sand, so completely identical in all its characters with the sands of the present shores and shallows of the gulf, that its origin cannot well be mistaken. Occasionally it embraces extensive beds of a red earthy marl. For instance, this marl may be seen in great abundance where excavations have been made for constructing a road in the bank, near a hundred yards northwest of the steamboat landing at Houston. By chemical examination I find a sample of this marl to consist mainly of carbonate of lime, red oxide of iron and silex. I believe it may be found of incalculable value to the city of Houston, and to the whole country above alluded to, inasmuch as limestone is not therein known to occur.

This marl, if calcined after the manner of burning lime, will become converted into a very good quicklime, of a reddish brown color. Nothing can be more efficient as the calcareous ingredient in all kinds of mortar, for laying bricks, making underground water tanks, and plastering houses internally and externally, where the color is no objection to its use. In fact it might be universally substituted for white lime, the iron or coloring material having only the effect to render it more hard and enduring than it otherwise would be.

The arable soil in this region consists of a basis of marine sand, mixed superficially with vegetable mould, but destitute of lime or saline matter. The application of this marl could not fail of greatly increasing its productiveness; and as the marl appears to be pretty generally distributed over the country, it will no doubt hereafter prove a cheap and efficient means of ameliorating the soil.

In travelling north from Houston, immediately after crossing Spring Creek, thirty miles distant, we come into a region gently rolling. Here we meet with small, rounded diluvial pebbles interspersed in the soil, some of jasper, others of quartz, flint, hornblende, etc. In my excursion, which extended as far north as the Mustang prairie, eighteen miles above Robbin's ferry, on the Trinity, these pebbles constantly presented themselves in the hills, but I saw none of greater size than a pigeon's egg. They are perhaps indicative of the nature of the rocks in the mountainous districts lying to the northwest, having been transported and worn by ancient marine currents. The eminences of the rolling region rise from one hundred to three hundred feet above the valleys. It is evidently of more ancient formation than the level region just described; nevertheless its outline or contour is most obviously the same as when the ocean left it, excepting the narrow gorges usually from ten to thirty feet deep, occupied by the present fresh water streams. Many of the high rolling prairies have their surfaces, especially their southern declivities, curiously marked with ridges and furrows five or six feet broad, as though they had been rudely tilled by some former race of giant plowmen. They have received the euphonical appellation of *hogwallow* prairies. Those who have observed the small, regular ripple marks, impressed by the waves on the sands of a shallow bay, or seen fluted and indented sandstone strata, high, dry, hard and a thousand miles inland perhaps,—the petrified ripple marks of an ancient sea, will have a correct idea in miniature of the appearance in question. May not these ridges and depressions in the sandy soil be the remains of successive ridges thrown up by the waves of a former sea? If not, whence came they?

Still more ancient than the beds of diluvial sand and pebbles, a formation of sandstone here and there presents itself at the surface, yet obviously underlying the whole of the rolling country. In some situations it has the hardness and all the other good qualities of a freestone, most valuable for building purposes. In other places it passes insensibly to the condition of incoherent sand. Often on the banks of the Trinity, the indurated sandstone alternates with mere sand beds. Several valuable quarries of freestone on and near the Trinity river, as at New Cincinnati and at the site of Osceola above the mouth of Bidais creek. At the latter place a most excellent building stone, of a light gray color and homogeneous texture, coming out in large oblong blocks, presents itself in a high bluff overhanging the river, in quantity apparently inexhaustible. The strata dip very slightly to the northwest, perhaps one foot in thirty. As large flat-boats may easily be laden with the freestone, and floated securely down the Trinity, it is not improbable that the future city of Galveston may be largely indebted to these and other neighboring localities for the materials of construction. In reference to its geological age, I may here remark, that between the limits of high and low water mark of the Trinity, this formation embraces extensive beds of lignite or brown coal, in which the woody structure is obvious. In places, huge logs and branches of opalized wood also occur, as do likewise the imperfect impressions of minor cauline plants. Those conversant with geology will therefore perceive that the Trinity sandstone belongs to the Tertiary series of formations.

The Trinity country, when its resources are developed, will in my opinion prove to be especially rich in salt springs. Only two localities came under my inspection; one of these is on Mr. Young's plantation, east side of the Trinity, near two miles higher than the mouth of Bidais creek; the other, similar to it but apparently more extensive, is on the Salinilla creek, a branch of Salt creek, west side of the Trinity, some four or five miles higher up, on the lands belonging to Dr. F. B. Page. Here are the unerring indications of an extensive subterranean deposit of salt or saliferous strata. Many acres of sand are here so constantly imbued with the saline transudation from below, as to be partially denuded of the usual vegetation, presenting here and there those succulent plants peculiar to salt marshes and other saline districts. If in some places a depression of a foot or so be made in the sand, a strong brine soon collects therein. I tasted of water thus procured, and it seemed to me to be nearly saturated with salt. I regret I had not the means of ascertaining its exact strength. I have visited several salt works in Ohio, Pennsylvania and New York. This salt water I think was more concentrated than even the Salina water; and no doubt wells of a moderate depth would command an inexhaustible supply of brine. The manufacture of salt in this place would of course prove exceedingly lucrative; for the country is well wooded, and the river banks, less than a mile distant, contain vast quantities of brown coal. The concentration might be accomplished with wood, or by solar evaporation upon the spot, or the water might be conveyed in a leaden tube to the banks of the Trinity, where the brown coal might easily be quarried out for supplying the furnaces with fuel. The salt could then, with very little expense, be floated down the river to Galveston, where it must always bear a fair price.

This region is well supplied with perennial springs, many of them seemingly pure as the water which falls from the clouds near the close of a rain storm, and many others imbued with diverse mineral qualities. A variety of mineral springs occur near Carolina and New Cincinnati; generally in wild ravines or embowered in picturesque groves; but those which I most particularly examined, rise on the Salinilla creek in an elevated and beautiful situation, in the midst of the singular saline prairie before mentioned, with the forest clad banks of the Trinity, half a mile or so to the east, and a noble prospect of high rolling prairie lawn and woodland, bearing away to the southwest. The Salinilla carbonated spring is sufficiently copious to yield nearly half a barrel per minute. Though its temperature as carefully ascertained is but 68° Fahr., yet it has some claim to be called a boiling spring, on account of the incessant bubbling up of nitrogen and carbonic acid gasses; with which latter, the water itself is strongly impregnated. I find the specific gravity of this water at 60° Fahr., to be 1.00 67. Four parts in a thousand by weight are saline mineral matter which can be obtained by evaporation. By means of numerous careful experiments with chemical reagents, I find the water to contain the following ingredients, viz. carbonic acid, chlorine, iodine, soda, lime, magnesia, organic matter a trace. We may therefore infer, that the gaseous and mineral contents of the spring are,

Carbonic acid,
Nitrogen,
Muriate of soda,
Muriate of magnesia,
Bicarbonate of soda,
Bicarbonate of lime,
Hydriodate of soda.

Experiments indicate the absence of sulphuric acid, iron and potash. The presence of iodine confers upon this water medicinal qualities of a most valuable kind. The same element has been detected in the Saratoga water, New York. The taste of the Salinilla water is unusually grateful and pleasant.

Near a quarter of a mile lower down the Salinilla creek, is a small sulphur spring, in the water of which I detected

Sulphuretted hydrogen,
Carbonic acid,
Muriate of soda,
Muriate of magnesia,
Bicarbonate of iron, mere traces
Silica, mere traces.

The dark sediment which subsides from this water upon standing, is mainly sulphuret of iron. The specific gravity of the water is 1.00 66. By evaporating 1000 grains of water, a saline residue is obtained weighing 2 $^3/_{10}$ grains.

Embellished as this site is with the most beautiful of Texas scenery, it may some day become a place of fashionable resort. To me the whole seemed like a landscape garden. The prairies everywhere presented a bewildering variety of flowers, rare, beautiful and nameless. Deer, and wild turkeys are numerous in the surrounding solitudes, and the clear lakes a few miles to the south abound with fine large fish.

In the banks of the Trinity, I often noticed deposits of a reddish brown iron stone, apparently a good iron ore; but my investigations respecting its extent or abundance, and the facilities which it might offer for the manufacture of iron, were not such as to allow me to speak decisively. Most of the small fountains which issue at frequent intervals from the steep banks of the river, above the brown coal formations, are strongly tinctured with iron, a circumstance which would seem to indicate abundance of iron ore.

This whole region abound to an extent perhaps unexampled in silicified or opalized wood: —wood changed to stone. Small oblong pieces are constantly met with on the higher portions of land, while in the banks of the Trinity, associated with the iron ore, and overlying the brown coal whole trees and fragments of trees, piled sometimes on upon another, present themselves completely transformed to stone. In some logs a diversified metamorphosis is observable: one portion of the vegetable structure having been replaced with silex, another with brown oxide of iron, and a third is bitumenized or converted to coal.

In concert with Dr. F. B. Page, I took considerable pains in the exploration of the Trinity brown coal formation. As no excavations for working have yet been made, the best places for inspecting the formation, are where the Trinity cuts its way through the high lands, or where its banks present themselves in bold high bluffs, as at New Cincinnati, and near the site of the projected town of Osceola. The coal lies in horizontal strata, dipping about one foot in thirty to the northwest. The main stratum at the latter place, just above Bidais creek, is represented by the concurrent statements of W. C. Brookfield, surveyor, Mr. James S. Hunter of Huntsville, Texas, Dr. Page, and some other persons whom I consulted, as between six and seven feet in thickness, the lower portion being three or four feet above low water mark. Unfortunately, during my sojourn there, the river was unusually high and turbid for the season of the year; I could not consequently verify the same by personal observations and measurements. The most considerable coal beds which I had opportunity fully to inspect, were in the Trinity bluffs, southwest side, at New

Cincinnati, six miles lower down, and just below the mouth of Salt creek, near six miles above. The workable stratum of brown coal in each of these localities is about five feet thick, and situated some fifteen feet or so above low water mark. In quality it is said to be precisely similar to the coal of the seven feet bed.

Specimens of average quality which I took from the bed near the mouth of Salt creek, have a specific gravity of 1.326. The proportion of carbon or coke, is forty seven parts in one hundred ($47/100$). The volatile portion consists of bitumen, creosote, pyroligneous acid and water. Upon burning 100 parts of the coal, there remains a trifle more than one part by weight of white ashes. The color of this coal is a dark umber brown, nearly black. Its ligniform structure is almost always easily discernible. It is readily ignited, burns with a pleasant flame, and with almost the same facility as charcoal. Although it has much less bitumen in its composition than the Pittsburgh or cannal-coal, it will yet prove valuable for nearly all purposes to which coal is applied; such as parlor use, the reduction of ore, and the generation of steam power. It is however ill adapted for the manufacture of inflammable gas.

This sort of coal is denominated *brown coal* or *brown lignite* by mineralogists. Sometimes it is called Bovey coal, because a thick bed of it has long been wrought at Bovey near Exeter in England. It occurs in many parts of the world, in some places in vast abundance, but generally in beds of far less extent than those of the Trinity. It is worthy of remark, that iron pyrites commonly so abundant and detrimental in coal, is here unusually scarce.

In estimating the value of these beds of lignite, it must be remembered, that the Trinity is a navigable stream, and almost the only one in Texas, which at this time deserves to be so ranked: that the city of Galveston, now with a population of 2000 and rapidly increasing, is situated on an island virtually destitute of timber. Hence Galveston needs fuel, and nothing is more probable than that the Trinity country will supply her. Each steamship plying between Galveston and New Orleans, consumes during the voyage both ways near one thousand barrels of coal, which at present costs them in New Orleans, an average of seventy or eighty cents a barrel. During twelve months past, the steamship Columbia is said to have expended $25,000 for coal alone. It is by no means improbable that coal might be profitably furnished to steam vessels at Galveston, from the Trinity, at one third or even one fourth this cost. If the demand can be supplied, Galveston will be one of the best coal markets in the world; for besides the requirements of ordinary commerce, steamships of war cruising in the Gulf of Mexico, will always find it a convenient place to lay in fuel.

New Orleans, July 15, 1830

*American Journal of Science and Arts,* 37 (1839):211–17

# Index

Pedernales River, 65
Perez, Antonio, 66–67
Peter, Robert, 9
Philadelphia, Pennsylvania, 18
physiography. *See Texas,
physiographic regions of*
Pin Oak. *See* Quercus palustris
Pine, 40–41.
*Pinus australis* (Long-Leaved
Pine), 41
*Pinus variabilis,* 41
plants and plant life. *See Texas,
plants and plant life in*
*Platanus occidentalis* (Sycamore),
62, 77
*Populus deltoides. See* Populus
laevigata
*Populus laevigata* (*Populus deltoides;*
Cottonwood), 77
porphyry, 48
*Portulaca oleracea* (Purslane),
38, 51
Post Oak, 45, 58, 64, 66
Presidio de Rio Grande, 73
Prickly Pear Cactus. *See* Cactus
Ferox *Nutt*
*Ptelea trifoliata* (Hop Tree), 62
Purslane. *See* Portulaca oleracea

quail, 43
quartz, 48, 69, 70, 71, 72, 74,
100
*Quercus falcata. See* Quercus
triloba
*Quercus palustris* (Pin Oak), 62
*Quercus rubra* (Red Oak), 62
*Quercus triloba* (*Quercus falcata;*
Spanish Oak; Southern Red
Oak), 62
*Quercus virens* (*Quercus virginiana;*
Live Oak), 51, 58, 62, 64, 69
*Quercus virginiana. See* Quercus
virens

Rafinesque, Constantine S., 9
Rattlesnake-Master. *See*
Eryngium aquaticum
rawhide, 55–56

Red-Berried Moonseed. *See*
Wendlandia popufolia
Redbud. *See* Cercis canadensis
Red Cedar, 60
Red Oak. *See* Quercus rubra
Reese, George W., 45
Rensselaer, Stephen Van, 4
Rensselaer Polytechnic Institute,
xi, 4
*Rhamnus caroliniana*
(Indian-Cherry), 59
*Rhus aromatica* (Fragrant Sumac),
62
*Rhus copallina* (Dwarf Sumac), 62
Riddell, John Leonard: on aerial
navigation, 21–23; background
and early years of, 3; botanical
interests and studies of, 6–7,
9–10, 11, 12–13, 15, 16–18;
binocular microscope by, 20; in
Cincinnati, 11–15; death of, 28;
on disease causation; 14, 25–26;
education of, 3–4; geological
interests and studies of, 13–14,
15, 18–19; inventions by, 20; as
lecturer in science, 5–6;
medical interests and studies
of, 25–27; as melter and refiner
for U.S. Mint, 17; natural
philosophy interests and studies
of, 19–25; at Ohio Reformed
Medical College, 7–10; on
matter, 24; on preventive
medicine, 26–27; as postmaster
of New Orleans, 28; and
professionalization of science,
27–28; as professor of
chemistry, Medical
Department of the University
of Louisiana, 15–29; and San
Saba gold mine, xii, 57–58, 70;
science career of, xi, 16, 29;
scientific interests of, 5, 6–7, 9;
on theory of gravitation,
23–24; Texas travels by, xiii,
xiv. *See also* Texas
Riddell, Mary Bone, 14–15, 16;
death of, xiii

*A Long Ride in Texas* was composed into type on a Paragon Publishing system in eleven point Janson with two points of spacing between the lines. Janson was also selected for display. The book was designed by Jim Billingsley, typeset by Publications Development Company, printed offset by Thomson-Shore, Inc., and bound by John H. Dekker & Sons, Inc. The paper on which this book is printed carries acid-free characteristics for an effective life of at least three hundred years.

TEXAS A&M UNIVERSITY PRESS
COLLEGE STATION